EXPLORATORY FACTOR ANALYSIS

SERIES IN UNDERSTANDING STATISTICS

NATASHA BERETVAS Series Editor

SERIES IN UNDERSTANDING MEASUREMENT

NATASHA BERETVAS Series Editor

SERIES IN UNDERSTANDING QUALITATIVE RESEARCH

PATRICIA LEAVY Series Editor

Understanding Statistics

Exploratory Factor Analysis
Leandre R. Fabrigar and Duane T. Wegener

Understanding Measurement

Item Response Theory
Christine DeMars

Reliability
Patrick Meyer

Understanding Qualitative Research

Oral History
Patricia Leavy

Fundamentals of Qualitative Research
Johnny Saldaña

LEANDRE R. FABRIGAR AND
DUANE T. WEGENER

EXPLORATORY FACTOR ANALYSIS

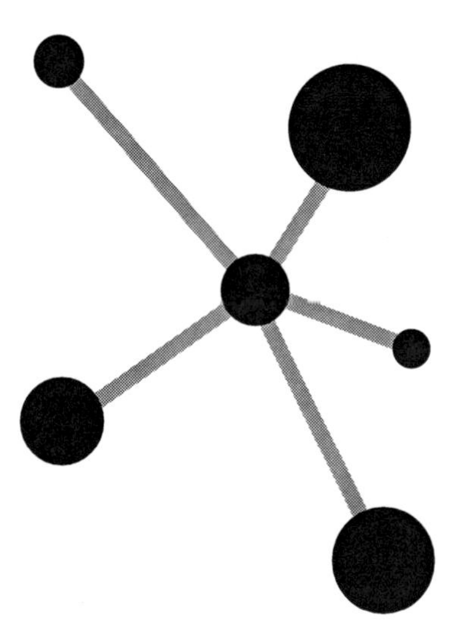

Oxford University Press, Inc., publishes works that further
Oxford University's objective of excellence
in research, scholarship, and education.

Oxford New York
Auckland Cape Town Dar es Salaam Hong Kong Karachi
Kuala Lumpur Madrid Melbourne Mexico City Nairobi
New Delhi Shanghai Taipei Toronto

With offices in
Argentina Austria Brazil Chile Czech Republic France Greece
Guatemala Hungary Italy Japan Poland Portugal Singapore
South Korea Switzerland Thailand Turkey Ukraine Vietnam

Published by Oxford University Press, Inc.
198 Madison Avenue, New York, New York 10016
www.oup.com

Library of Congress Cataloging-in-Publication Data
Fabrigar, Leandre R.
Exploratory factor analysis / Leandre R. Fabrigar and Duane T. Wegener.
p. cm. — (Understanding statistics)
ISBN 978-0-19-973417-7 (pbk. : alk. paper) 1. Factor analysis.
2. Psychology—Mathematical models. 3. Social sciences—Mathematical
models. I. Wegener, Duane Theodore. II. Title.
BF39.F23 2012
001.4′22—dc22 2011008725

Printed in the United States of America
on acid-free paper

ACKNOWLEDGMENTS

We would like to acknowledge the contributions to this book of two highly valued mentors and friends: Robert (Bud) C. MacCallum and Michael W. Browne. Bud and Michael first introduced us to factor analysis and other forms of latent variable modeling in the many graduate courses on quantitative methods that we took under their instruction. Throughout our careers, they have continued to influence our thinking regarding quantitative methods. This book reflects their views on factor analysis in many ways, some of which are directly acknowledged in citations of their many scholarly contributions to this research literature. However, their influence goes beyond the formal citations of their work. Many aspects of how we explain factor analysis (e.g., terminology and notation) and the recommendations we make regarding its implementation have been shaped by the courses we took under their instruction and the many conversations we have had with them about these issues over the years. Obviously, our acknowledgment of their important influences on our thinking should not be interpreted to suggest that they endorse everything stated in this book or that they are responsible for any errors that it may contain. Any mistakes are our own. However, to the extent this book successfully accomplishes its goals, Bud and Michael should receive much of the credit.

CONTENTS

EXPLORATORY FACTOR ANALYSIS

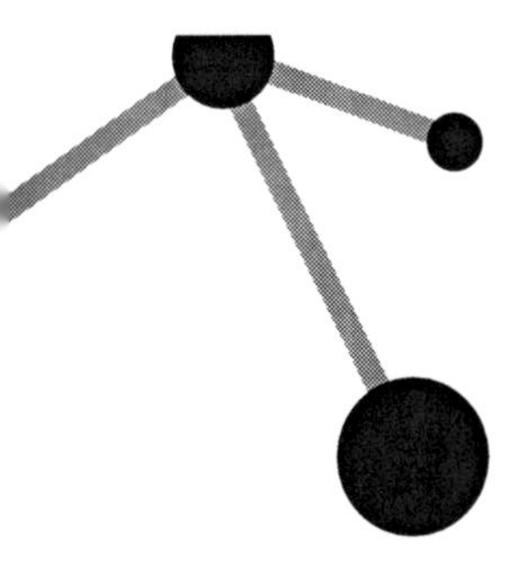

1

INTRODUCTORY FACTOR ANALYSIS CONCEPTS

The Goals of Factor Analysis

In the social sciences and many other disciplines, researchers often encounter a large set of observations or scores for a group of people (or objects). For example, a public opinion researcher might assess the attitudes of a sample of people on a wide range of sociopolitical issues. A marketing researcher might measure a number of beliefs and perceptions of consumers regarding a particular product. Or a psychologist might assess current emotional states of people by asking them to report the extent to which a variety of emotional descriptors reflect their current feelings.

In these and similar contexts, one question that often arises is whether this large set of scores can be more parsimoniously represented. That is, researchers often want to know what the underlying structure of associations is for a set of measures. Do these measures reflect a single underlying construct or do different subsets of measures represent a few distinct constructs? Or, on the other side of such questions, perhaps there is no parsimonious representation of the measures at all, and each measure reflects something uniquely its own that has little relation to any other measure in the set of observations.

Researchers can, of course, gain insight into these sorts of questions by simply examining the correlations among the measures. For instance, if all of the measures are effectively capturing the same underlying construct, we might expect the correlations among them to be very strong and of similar strength (as in Table 1.1).

Alternatively, if two distinct subsets of measures strongly reflect two different (and unrelated) constructs, we would expect measures comprising each subset to correlate very strongly with one another and to be largely unrelated to measures from the other subset (see Table 1.2).

Of course, in cases in which each measure assesses a distinctly different construct, we would expect a set of correlations in which none of the measures have substantial associations with one another (see Table 1.3).

Table 1.1

Hypothetical Correlations among a Set of Measures Assessing a Single Construct

	V1	**V2**	**V3**	**V4**	**V5**	**V6**
V1	1.00					
V2	.90	1.00				
V3	.90	.90	1.00			
V4	.90	.90	.90	1.00		
V5	.90	.90	.90	.90	1.00	
V6	.90	.90	.90	.90	.90	1.00

Table 1.2

Hypothetical Correlations among a Set of Measures Assessing Two Independent Constructs

	V1	**V2**	**V3**	**V4**	**V5**	**V6**
V1	1.00					
V2	.90	1.00				
V3	.90	.90	1.00			
V4	.00	.00	.00	1.00		
V5	.00	.00	.00	.90	1.00	
V6	.00	.00	.00	.90	.90	1.00

Table 1.3
Hypothetical Correlations among a Set of Measures Assessing Different Independent Constructs

	V1	V2	V3	V4	V5	V6
V1	1.00					
V2	.00	1.00				
V3	.00	.00	1.00			
V4	.00	.00	.00	1.00		
V5	.00	.00	.00	.00	1.00	
V6	.00	.00	.00	.00	.00	1.00

Unfortunately, understanding the structure underlying a set of measures is rarely as easy as understanding the data depicted in Tables 1.1–1.3. First, real data seldom produce patterns of correlations that are as clear and easy to discern as these hypothetical examples. Actual data will often show notable deviations from these patterns. Such deviations can make it difficult to gauge whether an observed pattern of correlations is sufficiently close to a hypothesized pattern to support the appropriateness of a particular structural representation of the data (e.g., all the measures assessing one construct or a certain small set of constructs). Equally problematic, simple visual inspection of correlations becomes increasingly difficult as the number of measures increases. The present example involves only six measures, which results in a relatively manageable matrix of 15 correlations. However, a set of only 20 measures produces a matrix with 190 correlations, and a set of 30 measures results in a matrix with 435 correlations. Thus, researchers routinely have sets of measures that are simply too large to understand with visual inspection of correlations.

Factor analysis was developed to overcome such challenges. Specifically, factor analysis refers to a set of statistical procedures designed to determine the number of distinct constructs needed to account for the pattern of correlations among a set of measures. Alternatively stated, factor analysis is used to determine the number of distinct constructs assessed by a set of measures. These unobservable constructs presumed to account for the structure of correlations among measures are referred to as *factors* or more precisely as *common factors*. The specific statistical procedures comprising factor analysis provide information about the number of common factors underlying a set of measures. They also provide information to aid in interpreting the nature of these factors. The nature of common factors is clarified by providing estimates of the strength and direction of influence each of the common factors exerts on each of the measures being examined. Such estimates of influence are usually referred to as *factor loadings*. For cases in which the researcher has no clear expectations or relatively incomplete expectations about the underlying structure of correlations, procedures exist to conduct *exploratory factor analysis* (EFA) or *unrestricted factor analysis*. In this book we focus on these procedures and refer to them as EFA. When a researcher has clear predictions about the number of common factors and the specific measures each common factor will influence, procedures are available to conduct *confirmatory factor analysis* (CFA) or *restricted factor analysis* (see Bollen, 1989).

A Conceptual Introduction to the Common Factor Model

The first factor analysis model was proposed by Charles Spearman (1904), and over the past century a number of other mathematical models have been suggested. However, most contemporary factor analysis procedures are based on L. L. Thurstone's (1935, 1947) *Multiple Factor Analysis Model*, which is now more often referred to as the *common factor model*. Although our primary goal in this book is to focus on the application of factor analysis to answer substantive research questions rather than to provide a detailed treatment of factor analytic theory, a very general understanding of the common factor model can be quite useful. By understanding the central tenets of this mathematical model, researchers can better appreciate the implications of some of their procedural choices when conducting factor analyses and can better interpret the meaning of their results. Importantly, the basic conceptual premises of the model are fairly intuitive and, to a large degree, can be understood without reference to extensive mathematical notations (although we will briefly touch upon the mathematical expression of the model later in this chapter).

Common Factors

The common factor model was proposed as a general mathematical framework for understanding or representing the structure of correlations among observed scores on a set of measures. In factor analytic terminology, these observed scores are usually referred to as *measured variables* (or alternatively as *manifest variables* or *surface attributes*). Thus, the term *measured variable* simply refers to any variable that can be directly measured (e.g., scores on a personality inventory, attitude measure, or academic skill test). The term *battery* is used to refer to a set of measured variables being examined in a specific factor analysis.

The central premise of the common factor model is that, when examining the correlations among measured variables, nonzero correlations will occur between measured variables because these variables are influenced (in a linear fashion) by one or more of the same unobservable constructs (or alternatively one might say these two variables are in part tapping on or measuring one or more of the same underlying constructs). For example, were we to administer a battery of academic skill tests to a group of people, we might observe a substantial correlation between scores on a vocabulary test and scores on a reading comprehension test. This correlation would be presumed to emerge because scores on both tests are influenced by the same unobservable construct (e.g., verbal ability). In other words, a person who has a high level of verbal ability would be expected to score high on both tests because verbal ability is important to performance on both tests. Likewise, a person with little verbal ability would be expected to score low on both tests because that person lacks a key skill needed to succeed on the two tests.

Thus, a *common factor* (also sometimes called a *latent variable* or *internal attribute*) is formally defined as an unobservable construct that exerts linear influences on more than one measured variable in a battery. It is referred to as "common" because it is common to more than one measured variable. The model proposes that, whenever correlations arise among measured variables, one or more common factors must exist within the battery. The model assumes that the number of common factors will be substantially less than the number of measured variables making up that battery. Indeed, were that not the case, there would be little point to conducting a factor analysis. The goal of factor analysis is to arrive at a relatively parsimonious representation of the structure of correlations. Such a representation is only parsimonious if the number of common factors needed to explain the correlations among measured variables is considerably less than the number of measured variables to be explained.

Unique Factors

The model further postulates the existence of *unique factors.* Unique factors are unobservable sources of linear influence on only a single measured variable in a battery. The model assumes that each measured variable is influenced by a unique factor. These unique factors represent that portion of the score on a measured variable that is not explained by the common factors. Because unique factors influence only a single measured variable in the model and are assumed to be unrelated to one another, unique factors cannot explain the existence of correlations among measured variables.

Conceptually, the model assumes that a unique factor can be further partitioned into two components: a *specific factor* and *error of measurement.* The specific factor refers to systematic sources of influence on a measured variable that are specific to only that measured variable. As such, specific factors are repeatable phenomena and do not harm the reliability of a measured variable. However, because researchers generally construct batteries with the intent that all the measured variables will reflect a single construct or that subsets of measured variables will reflect a single construct, the existence of strong specific factors is usually undesirable. That is, strong specific factors indicate that measures are strongly influenced by constructs they were not intended to assess. An example of a specific factor could be a bias in the wording of a self-report item that tends to encourage a particular response. The effects of the biased wording are systematic in that they may well affect the responses of a person consistently were the item completed at two points in time, but they are specific to that item in that the biased wording would not necessarily influence responses to other items in the battery.

The second component of unique factors is error of measurement. This term refers to random and, therefore, transitory influences on a single measured variable. Because these influences are random, the existence of a strong error of measurement component in unique factors will harm the reliability of a measured variable. An example of a potential source of random error might be an ambiguously worded self-report item. Depending on the context in which that item is presented and the current psychological state of the respondent, the same person might interpret the same item very differently at two points in time and, thus, provide very different responses.

Partitioning Variance in Measured Variables

Using the core premises of the common factor model, it is possible to conceptually partition the variance in a measured variable in several useful ways. First, the observed variance of a measured variable can be partitioned into:

observed variance = common variance + unique variance

The unique variance of a measured variable can be partitioned into:

unique variance = specific variance + error variance

In factor analyses, researchers also often refer to the common variance in terms of the proportion of the variance in the measure that is accounted for by the common factors. This proportion of variance accounted for by the common factors is referred to as the *communality*:

communality = common variance/observed variance

or

communality = 1 − (unique variance/observed variance)

Finally, in the context of the common factor model, the reliability of a measured variable can be expressed as:

reliability = (common variance + specific variance)/ observed variance

or

reliability = 1 − (error variance/observed variance)

A Graphical Depiction of the Common Factor Model

The common factor model can also be represented in the form of a diagram (often referred to as path diagrams in statistics). Figure 1.1 illustrates the common factor model in the context of our hypothetical example first provided in Table 1.2. In this example, the battery consisted of six measured variables and a pattern of correlations indicated the existence of two common factors. In the diagram representation of the common factor model, latent variables (common factors and unique factors) are represented as circles or ovals.

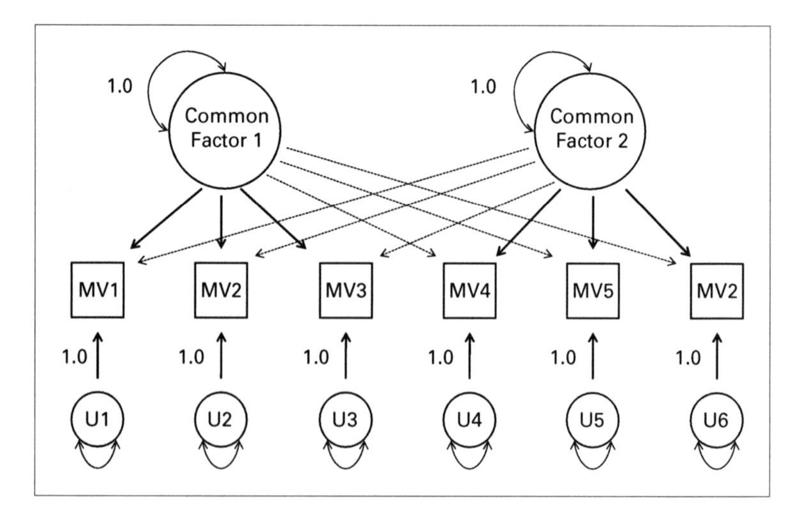

Figure 1.1. Graphical Representation of the Common Factor Model for an Example Involving Two Orthogonal Common Factors and Six Measured Variables

Measured variables are depicted with squares or rectangles. Linear causal effects are depicted with single-headed arrows indicating the direction of the influence. Linear nondirectional associations (e.g., correlations) are represented with double-headed arrows. A nondirectional association of a variable with itself represents the variance in that variable. Associations in the model (directional or nondirectional) that are assumed to have a specific numerical value are indicated by placing the assumed numerical value next to the association in the diagram.

In Figure 1.1, each measured variable is presumed to be influenced by a single unique factor. Each unique factor exerts an effect on one and only one measured variable. In common factor analyses, the values of these paths is assumed to be 1.0 (i.e., an increase of one unit in the unique factor is presumed to correspond to an increase of one unit in the measured variable). The double-headed arrow attached to each unique factor represents the variance of that unique factor. No assumptions are made regarding the value of each unique variance. Instead, these values are estimated from the data when the common factor model is fit to the data. The lack of any causal or nondirectional paths among the unique factors illustrates the assumption that unique factors are presumed to be independent of one another.

The figure further illustrates the existence of two common factors, with the first common factor strongly influencing the first three measured variables (i.e., MV1 – MV3; as represented by the solid directional arrows) but not appreciably influencing the remaining three measured variables (i.e., MV4 – MV6; as represented by the dashed directional arrows). Conversely, the second common factor is represented as having no appreciable influence on the first three measured variables but as exerting a strong influence on the last three measured variables. It is important to note that in EFA, no assumptions are made about the numerical values of the directional paths between common factors and measured variables. These values (i.e., the factor loadings) are estimated when the model is fit to the data. For this reason, the present diagram specifies possible paths between both common factors and all the measured variables. However, given the pattern of correlations presented in the hypothetical example in Table 1.2, one would expect the first common factor to only have an influence on the first three measured variables and the second common factor to only have an influence on the last three measured variables.

The lack of a double-headed arrow (i.e., a nondirectional association) between the common factors in the figure indicates that the two common factors are assumed to be unrelated (orthogonal) to one another. The assumption of orthogonal common factors is frequently made in EFA. In chapter 3, however, we will discuss how and why this assumption can ultimately be discarded at the rotation phase of EFA. The double-headed arrow associated with each common factor represents the variance of each common factor. In EFA, variances of common factors are usually assumed to be 1.0 and their means 0. These assumptions imply that common factors are in a standardized score metric (i.e., z-score metric).

The illustration of the common factor model in Figure 1.1 clearly implies that the first three measured variables should be highly correlated with one another because they are influenced by a common underlying construct. Likewise, the model also indicates that the last three measured variables should also be highly correlated with one another. In contrast, the model also indicates that none of the first three measured variables should have any sizable correlations with the last three measured variables. The relative lack of correlations is implied by the fact that there is no overlap in the influence of common factors between the first three and last three measured variables and by the fact that the two common factors are uncorrelated with one another.

Within the context of Figure 1.1 (which assumes the researcher had decided that two common factors are needed to account for the correlations), conducting a factor analysis ultimately produces several types of information when the model is fit to the data. First, the analysis provides the factor loadings, which are estimates of the strength and direction of the influence of the common factors on the measured variables. Second, the analysis produces estimates of the proportion of variance in each measured variable that is unique to that variable (or, conversely, the proportion of variance in each measured variable that is accounted for by the set of common factors). Finally, depending on the specific factor analytic procedures used, the analysis might also produce an estimate of how well the model fits the data (i.e., an estimate of how well the model accounts for the pattern of correlations among the measured variables).

A Simple Mathematical Introduction to the Common Factor Model

At this point, we have discussed the common factor model, first in terms of its basic conceptual premises, and then in the context of a pictorial representation of the model. It is also useful to consider the mathematical expression of the model. Having a working familiarity with the mathematical representation of the model can be very useful for understanding key features of some of the factor analytic procedures and the output from such an analysis. In presenting the model in its mathematical form, it is important to note that we will only be discussing one version of the model (i.e., the correlation structure model). We will not focus on computational issues related to the model, nor will we present mathematical proofs (for a more detailed mathematical treatment of the model, see Bollen, 1989; Gorsuch, 1983, Harman, 1976).

The common factor model is typically expressed in the form of matrix algebra. Matrices are rectangular representations used in algebra to present arrays of numbers or functions. In its matrix form, the correlation structure form of the common factor model (i.e., the expression of the model to account for the pattern of correlations among a battery of measured variables) is:

$$P = \Lambda\Phi\Lambda^{T} + D_{\Psi} \quad (1.1)$$

In the above equation, P refers to the population correlation matrix of measured variables in the battery of interest. Thus, using our example from Table 1.2 and the graphical representation of the model, this matrix would be a 6 x 6 correlation matrix among the measured variables reflecting the associations among these measured variables were the entire population of interest examined:

	MV1	MV2	MV3	MV4	MV5	MV6
MV1	1.00					
MV2	$\rho_{2\cdot1}$	1.00				
MV3	$\rho_{3\cdot1}$	$\rho_{3\cdot2}$	1.00			
MV4	$\rho_{4.1}$	$\rho_{4\cdot2}$	$\rho_{4\cdot3}$	1.00		
MV5	$\rho_{5.1}$	$\rho_{5\cdot2}$	$\rho_{5\cdot3}$	$\rho_{5\cdot4}$	1.00	
MV6	$\rho_{6.1}$	$\rho_{6\cdot2}$	$\rho_{6\cdot3}$	$\rho_{6\cdot4}$	$\rho_{6\cdot5}$	1.00

In the preceding matrix, ρ refers to a specific correlation between two of the measured variables in the population, with the first subscript used to refer to the row of the matrix and the second subscript to refer to the column of the matrix.

The matrix referred to as Λ represents the strength and direction of linear influence of the common factors on the measured variables. This matrix is often referred to as the *factor loading matrix*. In this matrix, columns represent the common factors and rows represent the measured variables. Using our example from earlier, the Λ would be:

	Common Factor 1	Common Factor 2
MV1	$\Lambda_{1 \cdot 1}$	$\Lambda_{1 \cdot 2}$
MV2	$\Lambda_{2 \cdot 1}$	$\Lambda_{2 \cdot 2}$
MV3	$\Lambda_{3 \cdot 1}$	$\Lambda_{3 \cdot 2}$
MV4	$\Lambda_{4 \cdot 1}$	$\Lambda_{4 \cdot 2}$
MV5	$\Lambda_{5 \cdot 1}$	$\Lambda_{5 \cdot 2}$
MV6	$\Lambda_{6 \cdot 1}$	$\Lambda_{6 \cdot 2}$

In the preceding matrix, $\Lambda_{1.1}$ refers to the factor loading representing the effect of the first common factor on the first measured variable. This value corresponds to the directional path between common factor 1 and MV1 in Figure 1.1. The element $\Lambda_{2.1}$ represents the factor loading indicating the influence of the common factor 1 on MV2, and so on.

The term Λ^T represents the transpose of the Λ matrix. A transpose is a mathematical operation in matrix algebra in which the rows of a matrix are re-expressed as the columns of a matrix. In the present example, Λ^T would be:

$$\begin{matrix} \Lambda_{1 \cdot 1} & \Lambda_{2 \cdot 1} & \Lambda_{3 \cdot 1} & \Lambda_{4 \cdot 1} & \Lambda_{5 \cdot 1} & \Lambda_{6 \cdot 1} \\ \Lambda_{1 \cdot 2} & \Lambda_{2 \cdot 2} & \Lambda_{3 \cdot 2} & \Lambda_{4 \cdot 2} & \Lambda_{5 \cdot 2} & \Lambda_{6 \cdot 2} \end{matrix}$$

The matrix Φ is the correlation matrix among the common factors. In the present example, this matrix would be as follows:

	CF1	CF2
CF1	1.00	
CF2	$\Phi_{2.1}$	1.00

As we noted earlier, in EFA, it is common to at least begin with the assumption that common factors are orthogonal (i.e., uncorrelated with one another). When this assumption is made, the Φ matrix can be omitted from the correlation structure equation:

$$P = \Lambda\Lambda^{T} + D_{\Psi} \tag{1.2}$$

The final matrix in Equations 1.1 and 1.2 is D_{Ψ}. This matrix is the covariance matrix among the unique factors. Thus, the diagonal elements of this matrix represent the variances of the unique factors and the off-diagonal elements represent the covariances among unique factors. Because the correlation structure model assumes that measured variables have been standardized, unique variances (diagonal elements) in the present equations actually correspond to the proportions of variance in each manifest variable attributable to unique factors. Additionally, because the model assumes that unique factors are orthogonal to one another, off-diagonal elements are assumed to be zero. In our example, the D_{Ψ} would appear as follows:

	U1	U2	U3	U4	U5	U6
U1	$D_{\Psi\,1\cdot1}$					
U2	0	$D_{\Psi\,2\cdot2}$				
U3	0	0	$D_{\Psi\,3\cdot3}$			
U4	0	0	0	$D_{\Psi\,4\cdot4}$		
U5	0	0	0	0	$D_{\Psi\,5\cdot5}$	
U6	0	0	0	0	0	$D_{\Psi\,6\cdot6}$

Table 1.4

Common Factor Model Represented in Full Matrix Form for Example

1.00							$\Lambda_{1\cdot1}$	$\Lambda_{1\cdot2}$		$\Lambda_{1\cdot1}$	$\Lambda_{2\cdot1}$	$\Lambda_{3\cdot1}$	$\Lambda_{4\cdot1}$	$\Lambda_{5\cdot1}$	$\Lambda_{6\cdot1}$		$D_{\Psi1\cdot1}$					
$\rho_{2\cdot1}$	1.00						$\Lambda_{2\cdot1}$	$\Lambda_{2\cdot2}$		$\Lambda_{1\cdot2}$	$\Lambda_{2\cdot2}$	$\Lambda_{2\cdot3}$	$\Lambda_{2\cdot4}$	$\Lambda_{2\cdot5}$	$\Lambda_{2\cdot6}$		0	$D_{\Psi2\cdot2}$				
$\rho_{3\cdot1}$	$\rho_{3\cdot2}$	1.00				=	$\Lambda_{3\cdot1}$	$\Lambda_{3\cdot2}$	×							×	0	0	$D_{\Psi3\cdot3}$		
$\rho_{4\cdot1}$	$\rho_{4\cdot2}$	$\rho_{4\cdot3}$	1.00				$\Lambda_{4\cdot1}$	$\Lambda_{4\cdot2}$									0	0	0	$D_{\Psi4\cdot4}$	
$\rho_{5\cdot1}$	$\rho_{5\cdot2}$	$\rho_{5\cdot3}$	$\rho_{5\cdot4}$	1.00			$\Lambda_{5\cdot1}$	$\Lambda_{5\cdot2}$									0	0	0	0	$D_{\Psi5\cdot5}$	
$\rho_{6\cdot1}$	$\rho_{6\cdot2}$	$\rho_{6\cdot3}$	$\rho_{6\cdot4}$	$\rho_{6\cdot5}$	1.00		$\Lambda_{6\cdot1}$	$\Lambda_{6\cdot2}$									0	0	0	0	0	$D_{\Psi6\cdot6}$

Considering the common factor model in it simplified form (i.e., assuming orthogonal common factors) as expressed in Equation 1.2, the full matrix expression of the model for our example is presented in Table 1.4. It can be useful to further illustrate the model by showing its expression with respect to specific elements of the population correlation matrix. The equation illustrated in Table 1.4 indicates that the elements of the population correlation matrix are a function of the matrix of factor loadings multiplied by the transpose of the factor loading matrix plus the covariance matrix of unique factors.

In matrix algebra, the multiplication of matrices involves multiplying elements of each row in the first matrix by the corresponding elements of each column in the second matrix. For instance, in the present example, the first element in row 1 of the factor loading matrix would be multiplied by the first element of column 1 of the transposed factor loading matrix. The second element of row 1 of the factor loading matrix would then be multiplied by the second element of column 1 of the transposed factor loading matrix, and so on. Multiplication of matrices thus produces a matrix with the same number of rows as the first matrix and the same number of columns as the second matrix (a 6 x 6 matrix in the present example). Addition of matrices simply involves adding each corresponding element of one matrix to the second matrix. Thus, addition of matrices can only be performed on matrices with the same number of rows and columns.

Taking these matrix algebra operations into account, the first diagonal element of the population correlation matrix, as indicated by the common factor model should be a function of:

$$\rho_{1.1} = 1.00 = \Lambda_{1.1}\Lambda_{1.1} + \Lambda_{1.2}\Lambda_{1.2} + D_{\Psi 1.1}$$

When considering the preceding equation, recall that all measured variables are assumed to be standardized. Thus, each measured variable has a variance of 1.00. The $\Lambda_{1.1}$ element represents the impact of common factor 1 on MV1. Squaring this term (as is implied by the first operation of the preceding equation) corresponds to the proportion of variance in MV1 explained by the first common factor. Similarly, $\Lambda_{1.2}$ reflects the impact of common factor 2 on MV1. Squaring this term indicates the proportion of variance in MV1 explained by the second common factor. The $D_{\Psi1.1}$ represents the proportion of variance in MV1 that is attributable to the unique factor. Taken together, the preceding equation implies that the total variance in MV1 will be the sum of the variance explained by common factor 1, common factor 2, and unique factor 1. More generally stated, the model implies that each measured variable will be a function of the sum of variance explained by each common factor in the model as well as the unique factor corresponding to that measured variable.

Let's now consider an off-diagonal element of the population correlation matrix:

$$\rho_{2.1} = \Lambda_{2.1}\Lambda_{1.1} + \Lambda_{2.2}\Lambda_{1.2}$$

The correlation between MV1 and MV2 will be equal to the product of the factor loadings of MV1 and MV2 on common factor 1 plus the product of the factor loadings of MV1 and MV2 on common factor 2. More intuitively stated, this equation states that the two measured variables would be expected to be highly correlated if common factor 1 strongly influences both measured variables and/or common factor 2 strongly influences both measured variables (i.e., there is overlap in the influence of common factors on the two measured variables). For instance, if MV1 and MV2 both had loadings on common factor 1 of .90 and loadings of .00 on common factor 2, the correlation predicted by the model would be: (.90 × .90) + (.00 × .00) = .81.

The model also implies that the measured variables would be expected to be uncorrelated if both common factors influenced neither measured variable or if the measured variables were influenced by different common factors (i.e., there is no overlap in the influence of common factors on measured variables). For instance, if MV1 had loadings of .90 and .00 on common factors 1 and 2, respectively, whereas MV2 had the opposite pattern of loadings, the predicted correlation between the two measured variables would be: (.00 × .90) + (.90 × .00) = .00. Note that there are no D_Ψ elements in the equation for the correlation between MV1 and MV2. This is because off-diagonal elements of D_Ψ are 0 and thus they have no impact on the correlations among measured variables. Thus, as can be seen from our calculations, the matrix equations illustrate the very same conceptual implications as did our verbal and graphical descriptions of the common factor model regarding when measured variables will and will not have strong correlations with one another.

Chapter Summary and Book Overview

Researchers often attempt to determine whether large sets of variables can be more parsimoniously represented as measures of one or a few underlying constructs. Factor analysis represents perhaps the most widely used set of statistical procedures for addressing this challenge. As we have seen, this set of procedures is based on a very general mathematical model, the common factor model. This model is the foundation of both EFA (the focus of the present book) and confirmatory factor analysis. The model advances a key set of premises regarding when and why measured variables are correlated with one another. These key premises of the model can be presented verbally, pictorially, or mathematically.

In application, researchers face several challenges in implementing the model. First, the researcher must determine if the common factor model is appropriate for their research context and if an exploratory implementation of this model is the desired approach to examining the model. That is, the researcher must decide if their research goals are adequately addressed by the model, if the data satisfy the assumptions of the model, and if the current (lack of) understanding of how the substantive measures behave suggests that an exploratory (rather than confirmatory) approach is advisable. As we discuss in chapters 2 and 4, there are a number of issues to be considered in addressing these questions.

Second, researchers must determine the specific version of the common factor model that will be most appropriate for a given battery of measured variables. That is, they must determine if the data are most appropriately represented by a single-factor version of the model, a two-factor version of the model, or so on. Chapter 3 outlines a number of procedures that methodologists have developed to aid in this decision.

The third major challenge researchers face is to determine the specific procedures they wish to use to estimate the parameters of the specific version of the common factor model that is selected. More than most statistical procedures, EFA offers researchers a range of different computational procedures for achieving the same basic set of objectives. Indeed, the number of choices available to researchers can at times seem overwhelming and, as a result, researchers are often unaware of the implications of choosing different procedures. As we note in chapters 3 and 4, the procedures differ in important ways, and the specific choice made can have important implications for what is found.

The final challenge that researchers face when implementing the model is interpreting the meaning of the results of the analysis. This task is obviously not a purely statistical concern and will be shaped by the researcher's substantive knowledge of the area of inquiry. That being said, there are certain basic considerations that can be applied to virtually any EFA. We will illustrate some of these principles in chapter 5.

References

Bollen, K. A. (1989). *Structural equations with latent variables.* New York: John Wiley.

Gorsuch, R. (1983). *Factor analysis* (2nd ed.). Hillsdale, NJ: Lawrence Erlbaum.

Harman, H. H. (1976). *Modern factor analysis* (3rd ed.). Chicago, IL: University of Chicago Press.

Spearman, C. (1904). General intelligence, objectively determined and measured. *American Journal of Psychology*, 15, 201–293.

Thurstone, L. L. (1935). *The vectors of mind.* Chicago, IL: University of Chicago Press.

Thurstone, L. L. (1947). *Multiple factor analysis.* Chicago, IL: University of Chicago Press.

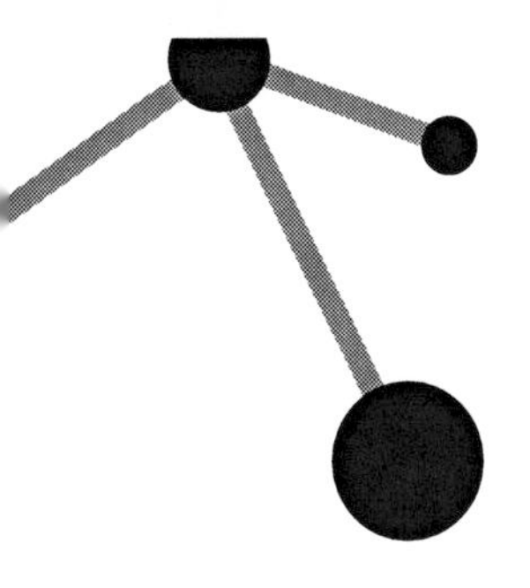

2

REQUIREMENTS FOR AND DECISIONS IN CHOOSING EXPLORATORY COMMON FACTOR ANALYSIS

IN CHAPTER 1, we provided an overview of the key conceptual premises underlying the common factor model. In the present chapter, we shift our attention to discussing key issues in determining when it is appropriate to conduct an exploratory analysis using this model. We begin our discussion of requirements for conducting exploratory common factor analysis by addressing the issue of what sorts of research questions are best explored by this type of factor analysis and the nature of the data necessary to properly conduct an analysis of this type. We will then address the issue of when exploratory factor analysis (EFA) versus confirmatory factor analysis (CFA) is most appropriate. Finally, assuming that an exploratory approach is selected, the researcher must then address the issue of whether the analysis should be based on the common factor model or the (different but related) principal component model.

Is EFA Suitable for the Research Question?

When considering the use of EFA, the first fundamental (and perhaps most obvious) issue that a researcher must consider is whether EFA is suitable for answering the research question of interest. As we noted in chapter 1, in a very general sense, factor analysis is used as a means of arriving at a more parsimonious representation of the underlying structure of correlations among a set of measured variables. Of course, this statement prompts at least two questions. Is factor analysis the only procedure for modeling the structure of correlations, and precisely why might someone want to do this in the first place?

The answer to the first question is certainly no. There are a number of procedures that can be used to model the structure of correlations among measured variables (including the principal components model, structural equation models, multidimensional scaling, cluster analyses, and others (see Stevens, 2009; Tibachnick & Fidell, 2007; Wegener & Fabrigar, 2000 for overviews). Structural equation modeling (SEM; or covariance structure modeling) provides a very general mathematical framework for modeling correlation (or covariance) structures. Factor analysis constitutes a special case of SEM in which the structure of correlations is being modeled with the intent of identifying the number and nature of latent constructs needed to account for the pattern of correlations among the measured variables when there is no goal to test specific hypotheses regarding directional associations among these latent constructs. That is, we use factor analysis when we want to know how many constructs a set of measured variables is assessing and what these constructs might be, but we are not yet at a point at which we want to test specific hypotheses about how the constructs might be causally related.

Thus, the nature of factor analysis within the broader SEM system also speaks to the types of research questions that can be usefully addressed using factor analysis (and EFA in particular). Typically, there are at least two types of specific research questions that fall under the general domain that factor analyses can address. It is important to note that these two types of questions are closely intertwined. Thus, they constitute a difference in emphasis rather than a fundamental difference in approach or objectives.

Construct Identification

One of the primary uses of factor analysis is in helping to identify the key constructs needed to account for a particular area of inquiry. Frequently, in the early stages of an area of research, the basic constructs making up the domain of interest have yet to be definitively identified. This process of construct identification is often accomplished by use of intuition and theoretical reasoning on the part of researchers. Factor analysis provides a statistical method for empirically assisting in the process of construct identification, rather than exclusively relying on intuition and theory. Thus, after a researcher has carefully defined the domain of interest, the researcher may then set about constructing and/or finding measured variables that seem to satisfy the conceptual definition of the domain. The researcher can then obtain scores on the measured variables from a sample of people (or objects) and conduct a factor analysis on these scores to determine the number of latent constructs (common factors) comprising the domain of interest. By examining which particular measured variables seem to be influenced by the same common factors, the researcher can reach conclusions regarding the nature of the constructs (see Wegener & Fabrigar, 2004, for a discussion of the roles of conceptual definition and theoretical assumptions in construct identification and measure construction).

There are numerous examples in psychology and other disciplines in which factor analysis has played a key role in the identification of constructs. One of the best-known examples comes out of personality research and the development of the influential Big Five theory of personality (see John & Srivastava, 1999). Early on in personality trait research, it was recognized that there were a vast number of personality traits that might be potentially used to describe stable differences in people's behavior, cognition, and emotion. An obvious question that arose was whether this immense domain of personality descriptors might ultimately be more parsimoniously represented as capturing or representing a relatively small number of core personality dimensions. That is, were one to examine the correlations among this vast set of personality descriptors, one might find that these descriptors could be grouped into a small number of subsets of descriptors, each of which consisted of descriptors reflecting a common underlying construct. Thus, personality researchers such as Norman (1963) compiled extensive lists of personality terms (sometimes consisting of thousands of terms) from sources such as unabridged dictionaries and then had respondents complete rating scales of terms drawn from these lists. Using factor analyses, Norman and others ultimately concluded that the correlations among these terms could be effectively represented with only five underlying factors. Researchers then examined the specific traits comprising each factor to reach conclusions regarding the nature of these factors. Ultimately, these factors have come to be interpreted as representing the fundamental personality dimensions of openness, conscientiousness, extroversion, agreeableness, and neuroticism.

Osgood, Suci, and Tannenbaum's (1957) classic research on semantic space constitutes another well-known use of factor analysis to identify key constructs. Osgood and his colleagues were interested in understanding the connotative meaning that people associated with objects. That is, they wanted to identify the underlying dimensions of connotative meaning that people used to classify or describe objects. To explore this question, they compiled an extensive list of hundreds of bipolar adjective rating scales (e.g., good/bad, strong/weak, active/passive). They then had numerous samples of people rate dozens of different objects using these scales and examined the structure of correlations among these rating scales using factor analysis. They found that the structure of correlations was well explained by three underlying factors. An examination of the adjectives making up each factor led them to conclude that the factors represented evaluation (consisting of adjectives such as good/bad, wise/foolish), potency (consisting of adjectives such as strong/weak, large/small), and activity (consisting of adjectives such as fast/slow, active/passive).

Measurement Instrument Construction

The second major use of factor analysis is to assist in the development of measurement instruments to assess constructs (see Floyd & Widaman, 1995; Gorsuch, 1997). Any factor analysis that provides information about the number and nature of underlying constructs in an area of inquiry also provides information regarding which specific measured variables effectively capture each factor. Therefore, factor analysis is widely used in the development of measurement instruments. Specifically, factor analysis provides two very general types of information that can be useful in constructing measurement instruments.

First, factor analysis provides valuable information regarding scale dimensionality. Researchers often construct measurement instruments consisting of multiple items, all intended to measure a single underlying construct. It is frequently the case that researchers then examine the reliability of their scales (usually using Cronbach alpha) and conclude that their scales are unidimensional if reliability is high. Such a conclusion does not necessarily follow because it has been shown that reliability can be high even when scales are multidimensional in nature (see Cortina, 1993; John & Benet-Martínez, 2000). Conversely, sometimes researchers may formulate subscales of items intended to assess distinct constructs, but these subscales might or might not tap distinct constructs or the constructs might not group in the way expected by the researcher. Fortunately, factor analysis provides a clear method for testing the dimensionality of a set of items and determining which items appropriately belong together as part of the same scale or subscale.

Factor analysis also provides useful information regarding the psychometric properties of specific items. An item that is strongly influenced by a factor that also strongly influences other items intended to measure the same construct suggests that the item may be effectively capturing its intended construct. Items with weak factor loadings on a factor that strongly influences other items intended to measure the same thing are likely poor measures of the intended construct. Additionally, for situations in which different subsets of items are intended to measure different constructs, factor analysis can be used to evaluate the impact of all the constructs on each item. This information can be used to identify items that may not be pure measures of a given construct (i.e., items that are strongly influenced by more than one factor). Thus, subscales can be constructed by selecting only those items that are strongly influenced by one factor and omitting items that are substantially influenced by several factors.

Are the Data Suitable for Factor Analysis?

Assuming that factor analysis is appropriate for the research question of interest, the next issue to consider is whether the data are suitable for factor analysis. This issue is obviously best considered before the data are collected in the first place. As with any statistical method, the results produced by a factor analysis are only as good as the data from which they are derived. No statistical procedure can overcome poor design choices. Thus, whenever possible, researchers should consider the requirements of the analyses they plan to conduct prior to collecting the data. In the context of factor analysis, there are two very general design issues that are especially important to consider when designing a study or using previously collected data.

Properties of the Measured Variables

One key consideration is the characteristics of the measured variables to be analyzed. There are several properties of the measured variables that are especially important in the context of factor analysis. First, and perhaps most fundamental, the validity of any factor structure that emerges will ultimately be determined by the adequacy with which measured variables have been sampled from the area of interest. If the measured variables do not adequately represent the domain of inquiry, the strength of some common factors may be greatly underestimated (if measured variables representing that factor are underrepresented in a battery) or, in the extreme, they may fail to be uncovered at all (if measured variables representing that factor are completely omitted or only represented by a few measured variables in a large battery). Conversely, if measured variables are included that do not appropriately belong to the area of inquiry, the resulting factor structure may be distorted such that the solution may be difficult to interpret because of the noise produced by these irrelevant measured variables or, in extreme cases, spurious factors (i.e., factors from another domain of interest) might emerge. Thus, whenever conducting a factor analysis, the researcher should carefully define the area of inquiry and systematically consider the extent to which each measured variable satisfies the conceptual requirements of the area of inquiry and the degree to which the battery as a whole adequately samples the area of interest.

A second aspect of the measured variables that should be considered is how many measured variables should be included in the battery. The answer to this question is obviously to some degree related to the prior issue in that the broader the area of inquiry, the greater the number of measured variables that will be required. Beyond this general point, it is important to recognize that factor analysis procedures tend to perform better when factors are *overdetermined* (i.e., when each factor has multiple measured variables strongly influenced by that factor). Studies have suggested that at least three to five measured variables reflecting each common factor should be included, although even more is generally desirable (see MacCallum, Widaman, Zhang, & Hong, 1999). For this reason, when designing a study in which factor analysis will be used, it is useful for the researcher to consider the maximum number of common factors that might be expected and the general nature of these factors. The researcher should then plan to include more than five measured variables to represent each possible factor (in the event that some measured variables do not load on their expected factor).

Yet another consideration is the quality of the measured variables making up the battery. Methodological research indicates that all things being equal, factor analytic procedures function better when the communalities of measured variables are high. One cause of low communalities is the use of measured variables with high levels of random error. Thus, sound measurement practices should be used when designing or selecting measured variables for inclusion in a factor analysis (see John & Benet-Martínez, 2000; Wegener & Fabrigar, 2004, for additional discussion).

A final set of issues to consider is the scale of measurement and distributional properties of the measured variables. We will discuss these issues in greater detail in chapter 4. The common factor model assumes that common factors exert a linear influence on measured variables. The extent to which this assumption is met can be affected by the scale of measurement of the measured variables (i.e., ordinal and nominal scales of measurement will generally not meet assumptions of linearity). Thus, in general, factor analysis using the common factor model is only appropriate when the measured variables have interval level or quasi-interval level scales of measurement (see Floyd & Widaman, 1995). Beyond this assumption of linearity, some specific procedures for fitting the common factor model to the data (e.g., maximum likelihood estimation) also assume that measured variables have multivariate normal distributions.

Properties of the Sample

When implementing a factor analysis, the properties of the sample from which the scores of the measured variables have been obtained should also be considered. Perhaps the most obvious question with respect to the sample is how large the sample must be in order to conduct a factor analysis. Textbooks on factor analysis and multivariate statistics routinely report rules-of-thumb, usually based on a ratio of participants to measured variables. For instance, Gorsuch (1983) recommended a ratio of five participants for each measured variable and that the sample size should never be less than 100. Others have recommended ratios as high as ten to one (e.g., Everitt, 1975; Nunnally, 1978). However, it is important to understand that these commonly advocated guidelines never had strong theoretical or empirical foundations but, instead, were based largely on intuition. Subsequent research has revealed them to be highly flawed (see MacCallum et al., 1999; MacCallum, Widaman, Preacher, & Hong, 2001; Velicer & Fava, 1998).

The central problem with such guidelines is that the sample size required to obtain accurate results depends on a variety of properties of the data and the model being fit. Most notably, when communalities of the measured variables are high (an average of .70 or higher) and each factor is overdetermined (at least 3 to 5 measured variables with substantial loadings on each factor), good estimates can be obtained with comparatively small sample sizes. When the data have much less optimal properties, even very large samples may be inadequate. Thus, a simple rule-of-thumb, such as a particular ratio of participants to measured variables, can, in some cases, greatly exaggerate the necessary sample size and, in other cases, badly underestimate the required sample size. Fabrigar et al. (1999) have suggested that under optimal conditions (communalities of .70 or greater and 3 to 5 measured variables loading on each factor), a sample of 100 can be adequate. Although a larger sample size is always desirable, with very well conditioned data, reasonable results can sometimes even occur with substantially smaller samples (less than 50) (see Preacher & MacCallum, 2002; de Winter, Dodou, & Wieringa, 2009). Under moderately good conditions (communalities of .40 to .70 and at least 3 measured variables loading on each factor), a sample of at least 200 should suffice. Finally, under poor conditions (communalities lower than .40 and some factors with only two measured variables loading on them), samples of at least 400 might be necessary, although, in such situations, it may be that even very large samples are inadequate.[1]

In many cases, it will be difficult for a researcher to be fully certain if the sample size is adequate prior to conducting the analysis. Thus, careful choices in design should always be taken to enhance the likelihood that optimal conditions will be obtained. Likewise, when possible, researchers should probably plan on only moderately good conditions in the data given that optimal conditions may sometimes be difficult to achieve.

Another issue to consider with respect to the sample is the extent to which the sample is representative of the population of interest. Frequently, researchers use samples selected for convenience (e.g., students drawn from a departmental participant pool, clients making use of a local health care facility). The most common result of such selective sampling is that the sample will be more homogenous than the broader population from which it is drawn. For instance, a researcher using a university sample to study cognitive ability tests would likely obtain a range of scores on these tests that did not adequately represent people at the lower range of ability. Reduced variance on the measured variables will tend to attenuate the correlations among measured variables. Such attenuated correlations will, in turn, tend to attenuate the factor loadings and correlations among factors.

Thus, when possible, samples randomly drawn from the population of interest are desirable. However, in many cases, such samples will not be feasible. Convenience samples need not be a problem so long as the biases in the sample are not strongly related to the constructs of interest. Thus, although some university samples might not be appropriate for tests of particular cognitive abilities (e.g., a sample of this sort might be highly nonrepresentative of the population on cognitive abilities such as verbal ability), for other constructs, such as perceptual abilities, a student sample might not be appreciably different from the population as a whole. Hence, researchers should carefully consider the nature of their samples and their relations to the domains of interest. When there is reason to expect that biases may exist in the sample, the results of any factor analysis should always be interpreted with an awareness of how sample homogeneity might have contributed to the results.

A final property of the sample to consider is the issue of missing observations. In general, factor analysis procedures assume that the matrix of correlations among the measured variables is based on a sample for which all members of the sample have contributed a full set of observations (i.e., "listwise" deletion is conducted such that a member of the sample is completely omitted from the analysis if one or more observations are missing). For cases in which a member of the sample has a large number of missing observations, it is probably most sensible to exclude that individual from the analysis. For cases in which observations on only a few measured variables are missing, several procedures are available to researchers (see Gorsuch, 1983). Sometimes, the mean of that measured variable from the overall sample is substituted for the missing value. Alternatively, sometimes an individual is selected from the sample at random and that person's value on the measured variable is used as the estimate of the missing value. Other more complex regression-based procedures for estimating missing data have been developed (e.g., Timm, 1970). These procedures can provide somewhat more accurate results than the simpler methods and function reasonably well in contexts in which up to 20 percent of the data are missing. More recently, even more sophisticated methods for estimating missing data such as maximum likelihood methods of data imputation have been proposed (e.g., see Arbuckle, 1996; Figueredo, McKnight, McKnight, & Sidani, 2000).

Is an Exploratory or Confirmatory Approach Most Appropriate?

Once the researcher has determined that the research question of interest can be addressed by factor analysis and that the data are appropriate for a factor analysis, the next choice that a researcher must make is whether an exploratory or confirmatory approach is most appropriate. Obviously, when the researcher has no expectations about the number of common factors and which measured variables will be influenced by the same common factors, an exploratory approach is advisable. In contrast, when the researcher has a theory that clearly specifies a precise number of factors and exactly which measured variables each factor should influence, a confirmatory approach is generally preferred.

Of course, in many situations, the precision of a researcher's expectations will fall somewhere between these two extremes. For instance, the researcher may have a very general idea about how many factors might emerge and some expectations about which measured variables might be influenced by these factors. However, the theory and/or prior data supporting these expectations might not be sufficiently developed to specify with a high level of confidence the exact number of factors and to make predictions about how each measured variable in the battery will be influenced by the factors. Similarly, when generating new items to measure a particular construct in a given domain, the researcher might not be sure that the new items will be influenced only by the construct of interest and not by related constructs in the domain. In such cases, we would also advocate an exploratory approach with the proviso that a confirmatory approach might be adopted later in the research program after exploratory analyses helped to more fully develop the researcher's hypotheses. Hence, a researcher might use an exploratory analysis in a first study and then in a second study use a confirmatory approach. Likewise, with a very large data set, a researcher might conduct an exploratory analysis on one half of the data and a confirmatory analysis on the other half.

In some cases, the researcher might not be able to confidently identify a single model as a preferred model, but might have a sufficient theoretical foundation to precisely specify two or three competing models. As long as there are only a few competing models and the theories motivating the models are adequately developed so that the two or three competing models can be fully specified (i.e., the exact number of factors can be specified and the measured variables each factor is expected to influence can be specified), a CFA is usually preferable to an EFA. However, whenever the number of competing models becomes comparatively large, an exploratory approach may be advisable because of the unwieldiness of fitting and comparing a large number of models. Moreover, when there exists a large number of plausible competing models, it may suggest that there is substantial ambiguity in the area of inquiry and that all potentially plausible models may not yet have been identified. An exploratory approach seems preferable in this context.

Regardless of whether an exploratory or confirmatory approach is adopted, it is important to recognize that differences between the approaches are more a function of emphasis than a fundamental difference in goals and underlying assumptions. Indeed, many of the presumed differences between EFA and CFA are more illusory than real. For instance, some researchers have argued that an advantage of CFA over EFA is the ability to formally quantify model fit, compare competing models with respect to their fit, and conduct statistical tests of parameter estimates (e.g., see John & Benet-Martínez, 2000). However, many of these perceived differences are not fundamental to the exploratory/confirmatory distinction but, rather, arise out of the particular model fitting procedures that have been commonly used in these two approaches. For instance, maximum likelihood parameter estimation has been far and away the most frequently used method of fitting models in CFA, but it has been much less commonly used in EFA. However, maximum likelihood parameter estimation can certainly be used in EFA, and, when so employed, it is possible to compute the same model fit indices commonly utilized in CFA. It is also possible to compute standard errors, confidence intervals, and statistical tests for model parameters when using maximum likelihood EFA (Cudeck & O'Dell, 1994). Therefore, these supposed benefits of CFA do not constitute strong reasons to shift from EFA to CFA, especially if existing understanding of the domain and measures is not sufficiently advanced to directly specify all the relevant alternative models in CFA.

The Common Factor Model or Principal Component Model?

Assuming that an exploratory approach is appropriate, the next decision that a researcher must make is whether to conduct the analysis based on the common factor model or based on the principal component model. There have been few issues in the factor analytic literature that have generated more debate among methodologists and produced more confusion among researchers than the common factor versus principal component decision. Systematic reviews of factor analytic studies have revealed that the majority of studies use principal component analysis (PCA) rather than common factor methods (e.g., Fabrigar et al., 1999). Researchers using PCA frequently assume that it is simply a form of EFA that produces results that are very similar to other forms of EFA. In point of fact, the first of these assumptions is not correct, and the second is highly debatable. PCA is based on a different underlying mathematical model than EFA, was originally designed for somewhat different goals, and in some cases can produce substantively different results.

As discussed in chapter 1, the common factor model was formulated as a general mathematical framework for understanding the structure of correlations among measured variables. It postulates that correlations among measured variables can be explained by a relatively small set of latent constructs (common factors), and that each measured variable is a linear combination of these underlying common factors and a unique factor (comprised of a specific factor and random error). PCA differs from the common factor model in several notable ways (Widaman, 2007; see also Thomson, 1939).

First, PCA was not originally designed to account for the structure of correlations among measured variables, but rather to reduce scores on a battery of measured variables to a smaller set of scores (i.e., principal components). This smaller set of scores are linear combinations of the original measured variable scores that retain as much information as possible (i.e., explain as much variance as possible) from the original measured variables. Thus, the primary goal of PCA is to account for the variances of measured variables rather than to explain the correlations (or covariances) among them. Similarly, PCA was not designed with the intent that the principal components should be interpreted as directly corresponding to meaningful latent constructs. Rather the components simply represent efficient methods of capturing information in the measured variables (regardless of whether those measured variables represent meaningful latent constructs).

Second, principal components are not mathematically or conceptually equivalent to common factors. Common factors are unobservable latent constructs that are presumed to cause the measured variables. The common factor model clearly distinguishes the influence of common factors on measured variables from the influence of unique factors on measured variables. Thus, the common factor model partitions the variance in measured variables into common variance and unique variance. Principal components, on the other hand, are directly constructed from the measured variables themselves and, as such, principal components contain both common and unique variance. Hence, the principal components model does not distinguish between common and unique variance. For this reason, some methodologists have argued that principal components analysis can be conceptualized as a special case of the common factor model in which unique variances are assumed to be 0 (e.g., see Gorsuch, 1990). We will further illustrate this viewpoint later in chapter 3 when we discuss methods of model fitting and contrast the computational approaches used in methods of fitting the common factor model in EFA from the computational approach used in PCA. In summary, PCA essentially involves a model in which a small set of principal components are constructed from the measured variables, and the ability of these components to predict the measured variables is assessed (as indexed by the principal component loadings).

Given that PCA and EFA are not the same, at least at the mathematical level, the questions then arise about whether these approaches are different at the practical level and if one approach can be regarded as generally superior to the other for purposes of construct identification and measurement instrument construction. This question has generated considerable debate among researchers. Some methodologists have argued that PCA is a viable substitute for EFA procedures and in some respects may be superior (e.g., see Velicer & Jackson, 1990a, 1990b). Advocates of PCA have typically based their position on several arguments. First, they note the greater computational simplicity of PCA and its less intensive computer memory requirements. Second, they argue that PCA generally produces very similar results to EFA. Third, they note the fact that EFA can sometimes produce Heywood cases (i.e., conceptually implausible or impossible estimates in which a communality is estimated to be 1 or greater than 1) whereas PCA does not produce these cases. Finally, advocates of principal components note that the components model is determinant insofar as it is possible to directly compute an individual person's score on a principal component whereas this cannot be done for common factors (which are unobservable latent variables).

Our own position is that these arguments are not especially compelling. In general, when the goal of research is to identify latent constructs for theory building or to create measurement instruments in which the researcher wishes to make the case that the resulting measurement instrument reflects a meaningful underlying construct, we argue that common factor (EFA) procedures are usually preferable. Our view is based on several counterarguments to the reasons put forth to prefer PCA and on arguments to prefer EFA that have been advanced by advocates of common factor procedures.

First, with respect to computational simplicity and computer memory requirements, this issue had some currency in the 1960s and 1970s. However, advances in computer hardware and software have rendered these differences trivial (Fabrigar et al., 1999). Even very large matrices of correlations can be analyzed with common factor methods in a matter of seconds.

Second, although PCA and EFA often produce similar results, EFA advocates have noted that there have been numerous documented cases of specific contexts in which the procedures produce substantive differences in results (e.g., Bentler & Kano, 1990; Fabrigar et al., 1999; Gorsuch, 1990; McArdle, 1990; Snook & Gorsuch, 1989; Tucker, Koopman, & Linn, 1969; Widaman, 1990, 1993, 2007). Differences are especially likely to emerge when communalities are comparatively low (.40 or less) and there are a modest number of measured variables loading on each factor (Widaman, 1993). These data characteristics are common in social science research. EFA advocates further note that when the data correspond to assumptions of the common factor model, EFA procedures produce more accurate estimates than PCA (McArdle, 1990; Snook & Gorsuch, 1989; Tucker et al., 1969; Widaman, 1990, 1993, 2007). Principal components, on the other hand, does not do appreciably better than common factor procedures when the data are consistent with the assumptions of PCA (e.g., the presence of little unique variance; see Gorsuch, 1990; McArdle, 1990; Velicer, Peacock, & Jackson, 1982).

With respect to the issue of Heywood cases, it is important to note that these problematic estimates often arise when the model has been seriously misspecified or the data severely violate assumptions of the common factor model (van Driel, 1978). For this reason, Heywood cases can be seen as having diagnostic value regarding potential problems with the model being fit or the data to which it is being fit (Velicer & Jackson, 1990a). PCA, on the other hand, does not solve these problems; it simply makes them less likely to be recognized (McArdle, 1990).

The indeterminancy of individual common factor scores is largely irrelevant to the primary goals of factor analysis. The goal of understanding the structure of correlations among measured variables does not require the computation of factor scores but merely the estimates of factor loadings and correlations among factors (McArdle, 1990). For typical cases in which factor scores might be of interest, scores are used to predict some other variable or serve as a dependent variable in an analysis. These sorts of questions, however, do not require the computation of factor scores (Fabrigar et al., 1999). Instead, they can be addressed in SEM. In SEM, common factors can be specified as correlates, predictors, or consequences of other variables in the model (either observed or latent), and estimates of the associations (directional or nondirectional) among variables can be obtained without computing factor scores.

Beyond these counterarguments, advocates of common factor procedures have also noted that EFA has several advantages over PCA. For example, they argue that the common factor model is based on more realistic assumptions because it directly acknowledges the existence of unique variance whereas PCA does not (Bentler & Kano, 1990; Loehlin, 1990). Given that most measured variables in real data will have at least some measurement error, it is unrealistic to assume unique variances are zero as is done in PCA.

Additionally, advocates of EFA point out that it is based on a testable model whereas PCA is not (Bentler & Kano, 1990; McArdle, 1990). That is, the common factor model specifies certain hypotheses about the data. The model can thus be fit to the data and rejected if fit is found to be poor. The principal components model on the other hand does not really provide a clearly testable set of hypotheses about the data, thus making it difficult to create meaningful indices of model fit that might be used to evaluate its performance.

It has also been noted that even when fit to the population, PCA produces parameter estimates (e.g., component loadings) that cannot be generalized beyond the specific battery from which they were derived, whereas EFA produces parameter estimates that can be generalized beyond the specific battery from which they are derived (Widaman, 2007). For example, the addition of a new measured variable to a battery alters PCA parameter estimates related to the original measured variables in the battery such as their loadings on the principal components. In contrast, presuming a newly added measured variable does not rely on new common factors that were not present in the original battery, the addition of that measured variable will not change EFA parameter estimates such as factor loadings for the original measured variables in the battery.

Summary

The primary theme of this chapter has been that there are a number of considerations that researchers should take into account before even undertaking an EFA. As we have seen, EFA is particularly well suited to two general types of research questions: construct identification and measurement instrument construction. Moreover, EFA is only appropriate when the data satisfy the underlying assumptions of the model (e.g., linear effects of common factors on measured variables) and the model fitting procedures used to estimate the parameters of the model (e.g., multivariate normality in the case of some model fitting procedures). We also noted that the current state of theorizing and empirical evidence available in the domain of interest are each important in determining whether an exploratory or confirmatory approach is most sensible. Finally, when an exploratory approach is preferred, researchers must decide if the analysis will be based on the principal component model or common factor model. Our own position is that common factor model procedures are generally superior to PCA when addressing the types of questions for which factor analytic procedures are typically used.

Note

1. Our discussion of sample size focuses on the sample size required for a model fitting procedure to produce parameter estimates (e.g., factor loadings) from a sample that closely approximate the true model parameter values in the population. This approach is not the only way in which sample size can be conceptualized. It can also be framed in terms of the required sample needed to reach a given level of statistical power to test a particular hypothesis regarding the model being fit to the data. We focus on the former rather than the latter conceptualization because inferential statistical tests are rarely a central focus of EFA. In contrast, inferential statistical tests are more often a central question of interest in the context of CFA or more general structural equation models. In such cases, procedures for determining the power of a given test in the context of factor analytic models and structural equation models have been proposed (e.g., see Kaplan, 1995; MacCallum, Browne, & Sugawara, 1996; Satorra & Saris, 1985; Saris & Satorra, 1993).

References

Arbuckle, J. L. (1996). Full information estimation in the presence of incomplete data. In G. A. Marcoulides & R. E. Schumacker (Eds.), *Advanced structural equation modeling: Issues and techniques* (pp. 243–277). Mahwah, NJ: Erlbaum.

Bentler, P. M., & Kano, Y. (1990). On the equivalence of factors and components. *Multivariate Behavioral Research, 25*, 67–74.

Cortina, J. M. (1993). What is coefficient alpha? An examination of theory and applications. *Journal of Applied Psychology, 78*, 98–104.

Cudeck, R., & O'Dell, L. L. (1994). Applications of standard error estimates in unrestricted factor analysis: Significance tests for factor loadings and correlations. *Psychological Bulletin, 115*, 475–487.

de Winter, J. C. F., Dodou, D., & Wieringa, P. A. (2009). Exploratory factor analysis with small sample sizes. *Multivariate Behavioral Research, 44*, 147–181.

Everitt, B. S. (1975). Multivariate analysis: The need for data and other problems. *British Journal of Psychiatry, 126*, 237–240.

Fabrigar, L. R., Wegener, D. T., MacCallum, R. C., & Strahan, E. J. (1999). Evaluating the use of exploratory factor analysis in psychological research. *Psychological Methods, 4*, 272–299.

Figueredo, A. J., McKnight, P. E., McKnight, K. M., & Sidani, S. (2000). Multivariate modeling of missing data within and across assessment waves. *Addiction, 95*(Supplement 3), S361–S380.

Floyd, F. J., & Widaman, K. F. (1995). Factor analysis in the development and refinement of clinical assessment instruments. *Psychological Assessment, 7*, 286–299.

Gorsuch, R. L. (1983). *Factor analysis* (2nd ed.). Hillsdale, NJ: Erlbaum.

Gorsuch, R. L. (1990). Common factor analysis versus principal components analysis: Some well and little known facts. *Multivariate Behavioral Research, 25*, 33–39.

John, O. P., & Benet-Martínez, V. (2000). Measurement: Reliability, construct validation, and scale construction. In H. T. Reis & C. M. Judd (Eds.), *Handbook of research methods in social and personality psychology* (pp. 339–369). New York, NY: Cambridge University Press.

John, O. P., & Srivastava, S. (1999). The big five trait taxonomy: History, measurement, and theoretical perspectives. In L. A. Pervin & O. P. John (Eds.), *Handbook of personality: Theory and research* (2nd ed., pp. 102-138). New York, NY: Guilford Press.

Kaplan, D. (1995). Statistical power in structural equation modeling. In R. H. Hoyle (Ed.), *Structural equation modeling: Concepts, issues, and applications* (pp. 100–117). Thousand Oaks, CA: Sage.

Loehlin, J. C. (1990). Component analysis versus common factor analysis: A case of disputed authorship. *Multivariate Behavioral Research, 25*, 29–31.

MacCallum, R. C., Widaman, K. F., Preacher, K. J., & Hong, S. (2001). Sample size in factor analysis: The role of model error. *Multivariate Behavioral Research, 36*, 611–637.

MacCallum, R. C., Widaman, K. F., Zhang, S., & Hong, S. (1999). Sample size in factor analysis. *Psychological Methods, 4*, 84–99.

McArdle, J. J. (1990). Principles versus principals of structural factor analyses. *Multivariate Behavioral Research, 25*, 81–87.

Norman, W. T. (1963). Toward an adequate taxonomy of personality attributes: Replicated factor structure in peer nomination personality ratings. *Journal of Abnormal and Social Psychology, 66*, 574–583.

Nunnally, J. C. (1978). *Psychometric theory* (2nd ed.). New York, NY: McGraw-Hill.

Osgood, C. E., Suci, G. J., & Tannenbaum, P. H. (1957). *The measurement of meaning*. Urbana, IL: University of Illinois Press.

Preacher, K. J., & MacCallum, R. C. (2002). Exploratory factor analysis in behavior genetics research: Factor recovery with small sample sizes. *Behavior Genetics, 32*, 153-161.

Sarris, W. E., & Satorra, A. (1993). Power evaluations in structural equation models. In K. A. Bollen & J. S. Long (Eds.), *Testing structural equation models* (pp. 181–204). Newbury Park, CA: Sage.

Satorra, A., & Sarris, W. E. (1985). The power of the likelihood ratio test in covariance structure analysis. *Psychometrika, 50*, 83–90.

Snook, S. C., & Gorsuch, R. L. (1989). Component analysis versus common factor analysis: A Monte Carlo study. *Psychological Bulletin, 106*, 148–154.

Stevens, J. P. (2009). *Applied multivariate statistics for the social sciences* (5th ed.). New York: Routledge.

Tabachnick, B. G., & Fidell, L. S. (2007). *Using multivariate statistics* (5th ed.). Boston, MA: Pearson.

Thomson, G. (1939). The factor analysis of ability. *British Journal of Psychology, 30*, 71–77.

Timm, N. H. (1970). The estimation of variance-covariance and correlation matrices from incomplete data. *Psychometrika, 35*, 417–437.

Tucker, L. R., Koopman, R. F., & Linn, R. L. (1969). Evaluation of factor analytic research procedures by means of simulated correlation matrices. *Psychometrika, 34*, 421–459.

van Driel, O. P. (1978). On various causes of improper solutions in maximum likelihood factor analysis. *Psychometrika, 43*, 225–243.

Velicer, W. F., & Fava, J. L. (1998). Effects of variable and subject sampling on factor pattern recovery. *Psychological Methods, 3*, 231–251.

Velicer, W. F., & Jackson, D. N. (1990a). Components analysis versus common factor analysis: Some issues in selecting an appropriate procedure. *Multivariate Behavioral Research, 25*, 1–28.

Velicer, W. F., & Jackson, D. N. (1990b). Components analysis versus common factor analysis: Some further observations. *Multivariate Behavioral Research, 25*, 97–114.

Velicer, W. F., Peacock, A. C., & Jackson, D. N. (1982). A comparison of component and factor patterns: A Monte Carlo approach. *Multivariate Behavioral Research, 17*, 371–388.

Wegener, D. T., & Fabrigar, L. R. (2000). Analysis and design for nonexperimental data: Addressing causal and noncausal hypotheses. In H. T. Reis & C. M. Judd (Eds.), *Handbook of research methods in social and personality psychology* (pp. 412–450). New York: Cambridge University Press.

Wegener, D. T., & Fabrigar, L. R. (2004). Constructing and evaluating quantitative measures for social psychological research: Conceptual challenges and methodological solutions. In C. Sansone, C. C. Morf, & A. T. Panter (Eds.), *The SAGE handbook of methods in social psychology* (pp. 145–172). New York: Sage.

Widaman, K. F. (1990). Bias in pattern loadings represented by common factor analysis and component analysis. *Multivariate Behavioral Research, 25*, 89–95.

Widaman, K. F. (1993). Common factor analysis versus principal component analysis: Differential bias in representing model parameters? *Multivariate Behavioral Research, 28*, 263–311.

Widaman, K. F. (2007). Common factors versus components: Principals and principles, errors and misconceptions. In R. Cudeck & R. C. MacCallum (Eds.), *Factor analysis at 100: Historical developments and future directions* (pp. 177–204). Mahwah, NJ: Erlbaum.

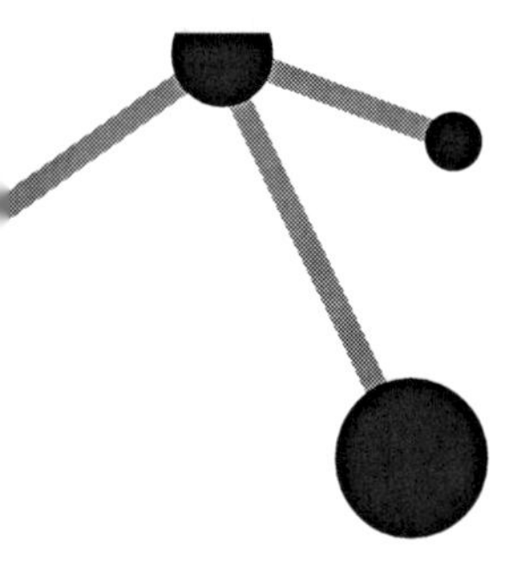

3

REQUIREMENTS AND DECISIONS FOR IMPLEMENTING EXPLORATORY COMMON FACTOR ANALYSIS

IN CHAPTER 2, we provided an overview of the key issues that should be considered before undertaking an EFA. In this chapter, we turn our focus to important requirements that must be satisfied and decisions that must be made in the actual implementation of an EFA. Few statistical procedures require more decisions on the part of a researcher and provide more choices in the implementation of the analysis. It is this aspect of factor analysis that users often find challenging and, at times, bewildering. The goal of this chapter is to provide readers with the necessary background to address each of these requirements/decisions in an informed manner. More specifically, we address three primary decisions that confront researchers when conducting EFA. First, assuming researchers have determined that an exploratory common factor model is appropriate for their research question and the data at hand, they must select from an array of model fitting procedures that estimate the parameters of the model. Second, researchers must determine how many common factors should be specified in the model when fitting it to the data. Finally, researchers must decide whether the resulting solution should be rotated to aid in interpretation of the results and if so, what specific rotation procedure should be used.

Choosing a Method of Fitting the Common Factor Model

Researchers must select one of several model fitting procedures (also sometimes referred to as factor extraction or parameter estimation procedures). Although numerous model fitting procedures have been developed, we will confine our discussion to the three most widely used methods: non-iterated principal axis (NIPA) factor analysis (also sometimes called principal factor analysis), iterated principal axis (IPA) factor analysis (or iterated principal factor analysis), and maximum likelihood (ML) factor analysis. In reviewing these procedures, our intent is not to provide a detailed discussion of the computational algorithms underlying these model fitting procedures. However, a general understanding of how the procedures arrive at their estimates is useful for interpreting the results produced and for understanding each method's strengths and limitations. Regardless of which procedure is selected, it is important to note that these procedures all assume the same underlying model (i.e., the common factor model). They merely differ in the computational approach by which the parameters of the common factor model are estimated. Thus, as might be expected, these procedures typically produce similar results. However, as we will discuss later, each approach has certain advantages and drawbacks, and there are some contexts in which they can produce different results.

NIPA Factor Analysis

You may recall from Chapter 1 that the correlation structure version of the common factor model (assuming all common factors are uncorrelated) is expressed by the following equation:

$$P = \Lambda\Lambda^{T} + D_{\Psi} \tag{1.2}$$

This equation expresses the common factor model in the hypothetical situation in which the common factor model holds perfectly in the population. The model can be alternatively and equivalently expressed by the following equation:

$$P - D_{\Psi} = \Lambda\Lambda^{T} \quad (3.1)$$

The challenge of any model fitting procedure is to arrive at estimates for the elements of each of these matrices. Unfortunately, in the present equation, we have three sets of unknowns: the matrix of measured variable correlations in the population, the matrix of unique variances, and the matrix of factor loadings.[1] Ultimately, for any model fitting procedure to be successful, the values for two of these matrices must be ascertained, and then the values for the remaining matrix can be calculated.

In considering this challenge, it is obviously the case that, in principle, the matrix of correlations among measured variables could be determined were one to obtain scores on the measured variables for all members of the population. Of course, in reality, this task can almost never be accomplished. However, a close approximation of the population correlation matrix can be obtained when one obtains scores on the measured variables for a sample drawn from the population. Thus, in practice, model fitting procedures attempt to derive estimates of the model that explain the correlations among measured variables in a sample. Therefore, NIPA factor analysis and other model fitting procedures attempt to solve for the following equation:

$$R \approx \Lambda\Lambda^{T} + D_{\Psi} \quad (3.2)$$

or alternative and equivalently:

$$R - D_{\Psi} \approx \Lambda\Lambda^{T} \quad (3.3)$$

where R represents the matrix of correlations among measured variables in a sample.

Several features of the preceding equations should be noted. First, in the Equations 3.2 and 3.3, the value of *R* is known in that it can be directly computed from the sample. Thus, one of the unknowns in the prior equations has been eliminated in Equations 3.2 and 3.3. Second, because the measured variable correlation matrix in the sample will almost never be a perfect representation of the measured variable correlation matrix in the population, we would not expect the common factor model to perfectly account for the correlations among measured variables in the sample, even if the model holds perfectly in the population. Of course, it is also true that in most real-world settings we would probably not expect the common factor model to hold perfectly in the population. At best, we would hope the model would closely approximate the population. Thus, in Equations 3.2 and 3.3, the = symbol has been replaced with the ≈ symbol denoting "approximately equal." Hence, when fitting the common factor model to the data, we will not expect to arrive at parameter estimates that perfectly account for the matrix of correlations among measured variables. The goal will instead be to arrive at parameter estimates that come as close as possible to accounting for the correlations among measured variables.

A feature of Equation 3.3 that also merits comment is the $(R - D_\Psi)$ term. Recall that the D_Ψ matrix is the covariance matrix among unique factors under the assumption that all unique factors are uncorrelated with one another. Thus, the diagonal elements are the unique variances associated with each measured variable and the off-diagonal elements are zero. When D_Ψ is subtracted from *R*, this results in a matrix in which the off-diagonal elements are the correlations among measured variables and the diagonal elements are the communalities of each of the measured variables (because subtracting a unique variance from the total variance of each measured variable, which is 1 when variables are standardized, produces the communality). This resulting matrix is referred to as the *reduced correlation matrix*. One of the major challenges in fitting the model for Equations 3.2 or 3.3 is to arrive at estimates of either the unique variances or equivalently the communalities. Once these have been determined, the values of the factor-loading matrix can then be calculated.

In the early days of factor analysis research, there was a great deal of debate regarding how the unique variances or communalities should be calculated. However, the most widely used approach and the approach that is employed in NIPA is to use the squared-multiple correlations (SMCs) of the measured variables as the estimates of the communalities. The SMC of each measured variable refers to the proportion of variance in that variable that is accounted for by the remaining measured variables in the battery. It thus reflects the amount of variance that the target variable has in common with other variables in the battery. These values can be directly calculated from the correlation matrix. Louis Guttmann (1956) noted that when the common factor model holds perfectly in the population, the SMC of a measured variable in the population is the lower bound for the communality of that measured variable in the population (i.e., the communality for that measured variable must be equal to or greater than the SMC). Based on this observation, Guttmann argued that the SMCs provided a reasonable approximation of the communalities and could be used as the basis for fitting the model.

In NIPA, the goal is thus to solve:

$$R - \mathrm{D}_{\Psi} \approx \Lambda\Lambda^{\mathrm{T}} \tag{3.3}$$

where $\mathrm{R} - \mathrm{D}_{\Psi}$ is the reduced correlation matrix in which the off-diagonal elements are the correlations among measured variables in the sample and the diagonal elements are the SMCs of the measured variables calculated from the sample. Thus, the only remaining unknown in the equation is the factor loadings (i.e., the elements of Λ). NIPA is a computational algorithm that produces the set of estimates for the factor loadings that comes closest to reproducing the reduced correlation matrix. That is, NIPA attempts to find the values for Λ that, when multiplied by its transpose, will produce a predicted reduced correlation matrix (frequently referred to as the *implied reduced correlation matrix*) that is as close as possible to the reduced correlation matrix observed in the sample.

We will not discuss the specifics of this computational algorithm. However, there are two aspects of the computational procedure that are useful for researchers to know. The first is how closeness between the matrix implied by the values of Λ and the reduced correlation matrix from the sample is mathematically defined. In the context of NIPA, this is done by computing the difference between each element of the implied matrix from its corresponding element in the reduced correlation matrix obtained from the sample. The matrix of these differences is referred to as the *residual matrix*. Overall discrepancy between the model and the reduced correlation matrix is quantified by squaring the values of the residual matrix and summing them. This residual sum of the squared differences (RSS) is referred to as the *discrepancy function*. The function is also referred to as the ordinary least squares (OLS) discrepancy function and is designated by the symbol F_{OLS}. The discrepancy function indicates the fit of the model to the data, with larger values reflecting worse fit. NIPA solves for the set of values for Λ that produces the smallest possible OLS discrepancy function.

A second issue to recognize is that any time the number of common factors in the model (or components in the case of principal components analysis) is equal to or greater than two, there will be more than one solution for the values of Λ that will produce an equally small discrepancy function value (i.e., the model is said to be indeterminant). Thus, some criterion must be specified for choosing one solution among these many equally good fitting solutions. In NIPA, this is done by setting restrictions (often referred to as identification conditions) on the solution for Λ that specify that this matrix must have certain mathematical relations to the eigenvalues and eigenvectors computed from the reduced correlation matrix.

The precise nature of these relations and the precise mathematical definition of eigenvalues and eigenvectors are not essential for researchers to understand when using NIPA (see Bollen, 1989; Gorsuch, 1983; Harman, 1976), although we will discuss some useful properties of eigenvalues throughout this chapter. The important point for readers to keep in mind is that, out of the vast array of equally fitting solutions that produce the smallest OLS discrepancy function, there will always be one and only one solution that satisfies the identification conditions set by NIPA. This permits the procedure to select a single solution and report those values. Additionally, it should also be noted that these identification conditions also guarantee that the solution selected will have several useful properties.

First, these conditions require that the first common factor (i.e., the first column of Λ) will be the common factor that accounts for the most variance in the measured variables. The second common factor (i.e., the second column of Λ) will be the common factor that is orthogonal to the first and accounts for the most remaining variance in the measured variables. The third common factor will be orthogonal to the first two factors and accounts for the most remaining variance and so on. A second resulting property of these identification conditions is that the largest eigenvalue from the reduced correlation matrix will be equal to the variance in the measured variables explained by the first common factor, the second largest eigenvalue will be equal to the variance explained by the second factor, and so on. Thus, in the context of EFA, eigenvalues can be thought of as numerical values that can be calculated from a reduced correlation matrix and correspond to the variance in measured variables accounted for by each of the common factors.

Once the solution for the factor loadings has been obtained, NIPA computes the final estimates of the communalities for the measured variables. The final communalities are computed by summing the squared loadings for each row of the factor loading matrix. Recall that when factors are restricted to be uncorrelated, the square of each factor loading corresponds to the proportion of variance in the measured variable accounted for by the factor. The final communality estimates are typically somewhat different from the initial estimates, which were based on the SMCs.

Another issue worth noting with respect to NIPA is that its resulting solution for Λ is generally only an intermediate step in conducting a factor analysis. Researchers rarely interpret this solution because this one solution among the many best fitting solutions for Λ has been selected solely for its computational ease rather than for its interpretability or conceptual plausibility. Thus, the solution for Λ that is usually interpreted is a solution that has been transformed to an equally good fitting, but more readily interpretable solution. Selection of a more interpretable solution is accomplished at the rotation phase. At rotation, the researcher can not only arrive at a more easily interpreted pattern of factor loadings but can also drop the restriction of orthogonal common factors. We will discuss rotation later in the chapter.

Before concluding our discussion of NIPA, it is useful to briefly contrast it with PCA. These two procedures share a number of computational similarities, but also differ in a significant way that helps to clarify differences between the common factor approach and the principal component approach. As we noted in the prior chapter, PCA does not differentiate between common and unique variance. Hence, the model does not include the D_ψ in its calculations. Instead, PCA involves solving for the following equation:

$$R \approx \Lambda\Lambda^{T} \tag{3.4}$$

Thus, PCA can be thought of as a special case of NIPA in which the unique variances of the measured variables are assumed to be zero (or alternatively and equivalently the communalities are assumed to be 1; Gorsuch, 1990). In summary, then, the primary computational distinction between NIPA and PCA is that NIPA involves finding the solution for Λ that best accounts for the reduced correlation matrix, whereas PCA involves finding the solution for Λ that best accounts for the original sample correlation matrix (i.e., the *unreduced correlation matrix*). In both procedures, the discrepancy function is the sum of the squared residuals. Likewise, identification conditions are similar for both procedures. Yet, because communalities are often far from one (and because the communalities often vary considerably across measures), it is not surprising that NIPA factor analysis and PCA can produce substantially different results in some settings.

IPA Factor Analysis

IPA is a variant of NIPA that is similar to it in most respects. As in NIPA, the first step in IPA is to compute the SMCs of the measured variables as the estimates of communalities so that the reduced correlation matrix can be obtained. IPA also uses the same method to obtain a solution for Λ. However, unlike NIPA, IPA does not terminate when the new estimates of the communalities are obtained by summing the squared loadings for each row of Λ. In IPA, once the new communalities have been computed, these new estimates are then used as the diagonal elements in R to produce a new reduced correlation matrix ($R - D_\psi$). A new solution for Λ based on this new reduced correlation matrix is then calculated, and this new solution for Λ is then used to compute yet another new set of communalities. These new communalities are then used in the reduced correlation matrix for the next cycle, and so on. Each cycle in the process is referred to as an *iteration*. This iterative process finally terminates when the procedure reaches a point at which the new estimates of the communalities obtained at the end of the iteration are nearly identical to those used as the initial estimates at the beginning of the iteration. When this occurs, the procedure is said to have *converged* on a solution.

ML Factor Analysis

Maximum likelihood is a model fitting procedure that many researchers frequently associate with CFA, but was actually originally developed for use in EFA (Lawley, 1940). Unlike NIPA and IPA, ML is based on two key assumptions about the data. First, it assumes that the data are based on a random sample drawn from some defined population. Second, it assumes that the measured variables have a multivariate normal distribution. This assumption implies that each measured variable has a normal distribution and that all associations between pairs of measured variables are linear.

Central to ML is the *likelihood function*. The likelihood function is a numerical value that reflects the relative likelihood of the observed data given a set of estimates for the model parameters. Simply stated, the goal of ML is to find that set of parameter estimates for a given model that are most likely to have produced the observed data. Stated another way, ML attempts to find the set of numerical values for the factor loadings and unique variances (or alternatively and equivalently the communalities) that will produce the largest possible likelihood function and, thus, are the estimates that, given that model, are maximally likely to have produced the data.

A detailed discussion of the computational procedure underlying ML is beyond the scope of the present chapter. However, it is useful to note several general features of how this procedure operates. First, rather than maximizing the likelihood function, it has been found that it is computationally easier to use an alternative function that is inversely related to the likelihood function. This alternative function is referred to as the *maximum likelihood discrepancy function* and is often designated with the F_{ML} symbol.

The ML discrepancy function is always equal to or greater than 0, with larger numbers indicating poorer model fit (i.e., a decreased likelihood that a model with a given set of parameter estimates could have produced the observed data). The ML discrepancy function will equal 0 if and only if the model fits the data perfectly. Thus, ML works to find the set of parameter estimates that produce the smallest possible ML discrepancy function.

A second important feature of ML is that it is an iterative procedure. ML begins with initial estimates of the communalities (usually using the SMCs), and an initial solution for the factor loadings is obtained. This initial set of factor loadings is calculated under identification conditions similar to NIPA and IPA so that a unique solution can be obtained. At each successive iteration, the ML algorithm calculates a new set of parameter estimates in an effort to further reduce the size of the ML discrepancy function. Thus, at each stage, the ML procedure attempts to find a set of parameter estimates that are more likely than the preceding set of estimates to have produced the observed data. The procedure converges on a final solution when the change in the value of the discrepancy function across iterations is smaller than some preset criterion.

After convergence, the procedure produces its final set of ML estimates of the factor loadings and unique variances (or, alternatively, the communalities). This solution is the set of estimates that, assuming the ML algorithm has functioned properly, are maximally likely to have produced the data given the model that has been fit. No other set of estimates for this model is more likely to have produced the data. However, just as in the prior fitting procedures, there are other sets of estimates that will fit the data as well. Thus, although no other set of estimates is more likely to have produced the data, there will be other sets of estimates that are equally likely to have produced the data. Therefore, as in NIPA and IPA, researchers seldom directly interpret the estimates directly produced by ML. Instead, the ML estimates are generally interpreted only after they have been rotated to a more readily interpretable solution.

Although ML produces similar information to NIPA and IPA, it also provides additional information. One such piece of information is the likelihood ratio test statistic (sometimes referred to as the χ^2 test of model fit). The likelihood ratio test statistic is produced by multiplying the final ML discrepancy function value by N - 1 (where N refers to the sample size upon which the correlation matrix is based). This resulting test statistic is approximately equal to a χ^2 with degrees of freedom equal to:

$$\text{df} = (p - m)^2 - (p + m)/2 \quad (3.5)$$

where p represents the number of measured variables in the correlation matrix and m represents the number of common factors in the model. The likelihood ratio test statistic is an index of the fit of the model and formally tests the null hypothesis that the common factor model with m factors holds perfectly in the population (i.e., the model and the data do not differ from one another). A significant test statistic indicates that the hypothesis has been rejected and thus the model does not hold perfectly in the population.

The likelihood ratio test statistic provides a seemingly appealing method for evaluating model fit in that it provides a clear criterion for acceptance or rejection of a model. However, in practice, the use of this index of model fit has serious limitations (see Browne & Cudeck, 1992; Cudeck & Henly, 1991; Hakstian, Rogers, & Cattell, 1982; Harris & Harris, 1971; Humphreys & Montanelli, 1975; MacCallum, 1990). First, many methodologists have argued that the test assesses a fundamentally unrealistic hypothesis: perfect model fit. In practice, we would, at best, expect a model to be a close approximation of reality, but seldom would we expect it to be perfect. Thus, rejection of perfect fit does not necessarily imply a model is not useful. The second problem with the likelihood ratio test is that it is very sensitive to sample size. As such, when sample sizes are extremely large, even very small ML discrepancy functions can produce a significant test statistic and lead to rejection of a very good performing model. In contrast, very small sample sizes can lead to a failure to reject the model, even when the ML discrepancy function is comparatively large, thereby indicating poor model performance.

For these reasons, methodologists have suggested that alternative indices of fit should be used when evaluating factor analytic models and structural equation models more generally. A number of such fit indices have been developed. These fit indices are often referred to as *descriptive fit indices*, because they attempt to quantify the magnitude of lack of fit between the model and the data rather than to reduce fit to a simple dichotomous judgment of perfect fit or lack of perfect fit. In principle, virtually any model fit index commonly used in CFA and SEM more generally can also be computed for an ML EFA model. Unfortunately, standard EFA programs such as those available in SPSS and SAS report few if any of these descriptive fit indices. Fortunately, some of these indices can be readily computed from the information provided by these programs (some of these indices are also available in specialized factor analysis programs such as CEFA; Browne, Cudeck, Tateneni, & Mels, 2010). We will discuss one such index in the current chapter: Root mean square error of approximation (RMSEA). In Chapter 5, we will discuss how this index can be computed from standard output provided by ML EFA procedures in SPSS and SAS.

RMSEA was originally proposed by Steiger and Lind (1980) and extensively discussed by Browne and Cudeck (1992). RMSEA is formally defined by the following equation:

$$\text{RMSEA} = \text{square root } (F_0/\text{df}) \qquad (3.6)$$

where F_0 refers to the estimate of the ML discrepancy function in the population, and df refers to the degrees of freedom for the model being fit to the data. The best estimate of the population ML discrepancy function is:

$$F_0 = F_s - (\text{df}/N - 1) \qquad (3.7)$$

where F_s is the ML discrepancy function computed for the model from the sample. RMSEA can be conceptualized as an index of discrepancy between the model and the data per degree of freedom of the model. In RMSEA, large values indicate poorer fit, and a value of 0 indicates perfect fit. Some methodologists (Browne & Cudeck, 1992; MacCallum, Browne, & Sugawara, 1996) have suggested that RMSEA be interpreted according to the following guidelines:

< .050	close fit
.050 to .080	acceptable fit
.081 to .100	marginal fit
>.100	poor fit

RMSEA has several properties that make it a particularly useful fit index. First, it has been found to perform relatively well in detecting misspecified models (e.g., see Hu & Bentler, 1998). Second, RMSEA takes into account model parsimony. That is, the fit index will only improve with a model of increasing complexity if the reduction in the discrepancy function is sufficient to offset the decrease in the degrees of freedom for the model. Thus, when comparing two models that produce similar discrepancy functions, RMSEA will favor the more parsimonious model. Finally, RMSEA has a known sampling distribution. This property permits the calculation of confidence intervals, thereby allowing researchers to determine the precision of the estimate of model fit. For example, consider the two following situations:

RMSEA = .06 (90% CI: .00 to .14)

RMSEA = .06 (90% CI: .05 to .07)

In both cases, our estimate suggests a model with acceptable fit. However, in the first case, the confidence interval indicates that the precision of this estimate is relatively low and, thus, in the population model, it could actually fit perfectly or the fit could be quite poor. In the second case, the model fit is also acceptable, but here the precision is such that we know the model at best fits well and at worse fits acceptably. Thus, computation of both the point estimate of RMSEA and its confidence interval is useful.

Beyond the RMSEA fit index per se, use of ML EFA enables one to compute standard errors, confidence intervals, and statistical tests for model parameters (Cudeck & O'Dell, 1994). Unfortunately, factor analysis procedures in standard programs such as SPSS and SAS do not provide all of this information. It is, however, available in specialized factor analysis programs such as CEFA.

Comparisons among Model Fitting Procedures

NIPA, IPA, and ML all fit the common factor model to the data. Thus, given that all three procedures are based on the same underlying model, one would both hope and expect these procedures to generally produce similar results. Under comparatively good conditions (e.g., strong common factors are present, the model is not seriously misspecified, and the data do not severely violate assumptions of the model or model fitting procedures), the procedures do produce very similar results (e.g., see Briggs & MacCallum, 2003; Widaman, 1993). However, as conditions become less optimal, some differences can be observed.

NIPA and IPA generally produce similar results, though IPA tends to produce slightly more accurate estimates (Widaman, 1993). Differences between the procedures are most pronounced under conditions of low communalities and low measured variable to common factor ratios. However, even here, the differences are comparatively small. One potential drawback of IPA is that iterative processes can sometimes produce *improper solutions* in that the procedure fails to converge on a final solution or produces a solution with Heywood cases (i.e., impossible outcomes, such as communalities greater than 1 or negative variances). However, IPA has been found to be fairly robust to such problems (e.g., Finch & West, 1997; Briggs & MacCallum, 2003), and, as noted earlier, the occurrence of such problems can sometimes have diagnostic value for assessing the quality of the model and problems with the data.

Comparing ML to NIPA and IPA, the greatest strength of ML is the fact that indices of model fit can be readily computed for this procedure and that it also permits computation of model parameter standard errors, confidence intervals, and significance tests. As we will discuss in the next section of this chapter, the ability to compute model fit indices in particular constitutes a significant practical advantage when evaluating models and judging whether an appropriate number of common factors has been specified in the model. That being said, ML also has some drawbacks. First, it assumes multivariate normality, whereas NIPA and IPA do not.[2]

Although ML is comparatively robust to violations of this assumption, it can nonetheless prove problematic when severe violations occur. In Chapter 4, we discuss procedures for determining if violations are sufficiently severe to warrant concern and what steps can be taken if this occurs. Another drawback of ML is that research has suggested that when moderate amounts of model error and/or sampling error are present, ML is less effective than IPA in recovering weak common factors (Briggs & MacCallum, 2003). Finally, ML is somewhat more prone to produce improper solutions than IPA (Finch & West, 1997; Briggs & MacCallum, 2003) and on occasions can fail to converge on a solution or terminate prematurely on a solution (i.e., encounter a local minimum in the discrepancy function). Fortunately, such problems tend to be rare unless the model is seriously misspecified or the data severely violate assumptions of the model or ML fitting procedure.

In general, because of the additional information provided by ML, ML is preferable to IPA and NIPA in most contexts. So long as non-normality is not severe and/or there is no reason to expect substantively important weak common factors, the additional information provided by ML will probably outweigh its drawbacks. That being said, even in these contexts, it is always prudent to examine the IPA solution as well as the ML solution to confirm that the procedures produce comparable results. Substantial differences between the procedures can serve to highlight the need for the researcher to further consider the model being fit and the properties of the data.

Determining the Appropriate Number of Common Factors

Once a researcher has chosen a method of fitting the common factor model, that researcher must then determine how many common factors will be specified in the model to be fit to the data. This decision has long been recognized as one of the great challenges facing researchers when conducting EFA, and, as such, a large methodological literature on this topic has accumulated over the years. Much of this literature has been directed toward the development of mechanical rules that will reliably indicate the true number of common factors (or principal components in the case of PCA) for a given battery of measured variables. Many methodologists consider such an approach to be misguided.

As noted earlier, whenever the common factor model is being fit to sample data, it is extremely unlikely that the model would fit perfectly. One reason for lack of fit of the model to a sample correlation matrix is *sampling error* (error resulting from the fact that the correlations computed from any sample are unlikely to perfectly correspond to the values of correlations among measured variables in the population). However, even if a population correlation matrix were available, it is not realistic to expect that any comparatively parsimonious common factor model (i.e., a model with substantially fewer common factors than measured variables) will fit perfectly. Lack of fit is likely to arise from phenomena such as nonlinear influences of the common factors on the measured variables and the influences of minor common factors that cannot be represented in a parsimonious model (MacCallum, Browne, & Cai, 2007; Tucker & MacCallum, 1991). Lack of fit arising from such model imperfections is referred to as *error of approximation*.

Recognizing that we will never expect a parsimonious model to be a perfect representation of the data and that any model will, at best, be a plausible simplified approximation of the data, the reasonable objective of factor analysis is not to arrive at the true or correct number of common factors (at least in the strict sense of these terms). Rather, the more sensible goal is to arrive at a useful or appropriate number of common factors for our model, where *useful* or *appropriate* is intended to reflect both statistical utility (i.e., the number of common factors needed to closely approximate the pattern of correlations among measured variables) and conceptual utility (i.e., a number of common factors that substantially simplifies our data and in which the nature of the common factors can be readily interpreted). Stated another way, our objective is to identify the number of major common factors underlying a battery of measured variables. Ideally, the number of major common factors is that number of common factors for a model in which: (1) the model does a good job accounting for the correlations among the measured variables, (2) a model with one fewer common factors would do substantially worse in accounting for the correlations, (3) a model in with one more common factor would not do appreciably better in explaining the correlations, and (4) all common factors in the model are readily interpretable and can be related to constructs of theoretical utility to the domain of interest.

These four criteria are relevant when evaluating the various procedures for determining the appropriate number of common factors (or principal components). As we will see, some of these procedures do rather poorly at satisfying any of the criteria. Others meaningfully address some, but not all the criteria. None of them address all the criteria or are infallible in addressing even those criteria to which they are relevant. Thus, determining the appropriate number of common factors is a decision that is best addressed in a holistic fashion by considering the configuration of evidence presented by several of the better performing procedures. Moreover, it is a decision that is as much theoretical as it is statistical.

Eigenvalues-Greater-than-One Rule

Probably the most widely used procedure for determining the number of factors is the eigenvalues-greater-than-one rule, also sometimes referred to as the *Kaiser criterion*. This procedure involves computing the eigenvalues from the unreduced correlation matrix or reduced correlation matrix and then simply examining the number of eigenvalues that are greater than one. The number of eigenvalues that exceed one is then used as the number of common factors or principal components to be specified in the model. This procedure is appealing in its simplicity and objectivity. Unfortunately, it also has serious flaws.

First, although the rule had an underlying logic in its application to eigenvalues from the unreduced correlation matrix used in PCA, it was never intended as a procedure for use with eigenvalues from the reduced correlation matrix (Guttman, 1954; Kaiser, 1960). Thus, the rule has no theoretical rationale in the context of common factor analysis and is an inappropriate criterion for eigenvalues from the reduced correlation matrix (Gorsuch, 1980; Horn, 1969). This fact presents researchers with a problematic choice when implementing this rule. A researcher could choose to use the unreduced matrix as the basis for implementing the rule and thus the criterion would have a conceptual foundation. However, in the context of EFA, this would create a logical contradiction. The researcher would in effect be using the unreduced matrix in Equation 3.3 to determine how many columns of the factor loading matrix in Equation 3.4 are required for it to account for the reduced matrix in Equation 3.4. Logically, it should be the sample matrix that the factor loadings are intended to explain that should be the basis for deciding how many columns the factor loading matrix requires. Stated another way, it should be the variance accounted for by the common factors that is used to determine how many common factors are required to explain the data rather than the variance accounted for by the principal components. Of course, conversely, if the goal is to determine the number of principal components required to account for data, it should be the variance accounted for by those principal components rather than the variance accounted for by common factors that is used as the basis for the decision.

A second problem with this rule is that, as with any purely mechanical rule, it can lead to arbitrary decisions with little conceptual basis. For instance, a principal component or common factor with an eigenvalue of 1.01 would be judged to be meaningful, whereas a principal component or common factor with an eigenvalue of .99 would not, despite the fact that the difference in variance accounted for by the two components/factors would be trivial. Moreover, there is nothing to suggest that components or factors satisfying this rule will be readily interpretable or will constitute a substantial improvement in the fit of the model.

Finally, numerous simulation studies have been conducted examining the efficacy of this procedure in the context of both PCA and EFA (Cattell & Jaspers, 1967; Cattell & Vogelmann, 1977; Hakstian, Rogers, & Cattell, 1982; Linn, 1968; Tucker et al., 1969; Zwick & Velicer, 1982, 1986). All these studies have indicated that the procedure performs poorly, often leading to substantial overfactoring (i.e., identifying more factors than were present in the population used to create the sample) and occasionally to underfactoring (i.e., identifying fewer factors than in the simulated population).

Scree Test

Probably the second most widely used method of determining the number of factors is the *scree test*, originally proposed by Cattell (Cattell, 1966; Cattell & Jaspers, 1967). This procedure involves constructing a graph in which eigenvalues from the unreduced or reduced correlation matrix are plotted in descending order. The resulting graph is then examined to determine the number of eigenvalues that precedes the last major drop. This number corresponds to the number of factors to be specified in the model. For example, consider the graph in figure 3.1. In this graph, the *last* major drop occurs following the third eigenvalue. Thus, this scree plot would suggest that a three-factor model is most appropriate.

When making use of the scree test, one issue that researchers are often not sufficiently sensitive to is which eigenvalues should be plotted. Eigenvalues from the sample correlation matrix (unreduced correlation matrix) or the eigenvalues from the reduced correlation matrix can be plotted. When conducting an EFA, as was explained in the context of the eigenvalues-greater-than-one rule, it is more sensible to plot the reduced matrix eigenvalues because these are the eigenvalues that more directly correspond to the extracted common factors. In contrast, if conducting a PCA, it would be more sensible to plot eigenvalues from the sample correlation matrix. Unfortunately, factor analysis procedures in standard statistical programs do not always plot the eigenvalues that correspond to the analysis being conducted (i.e., in some programs, the eigenvalues reported do not change depending on whether PCA

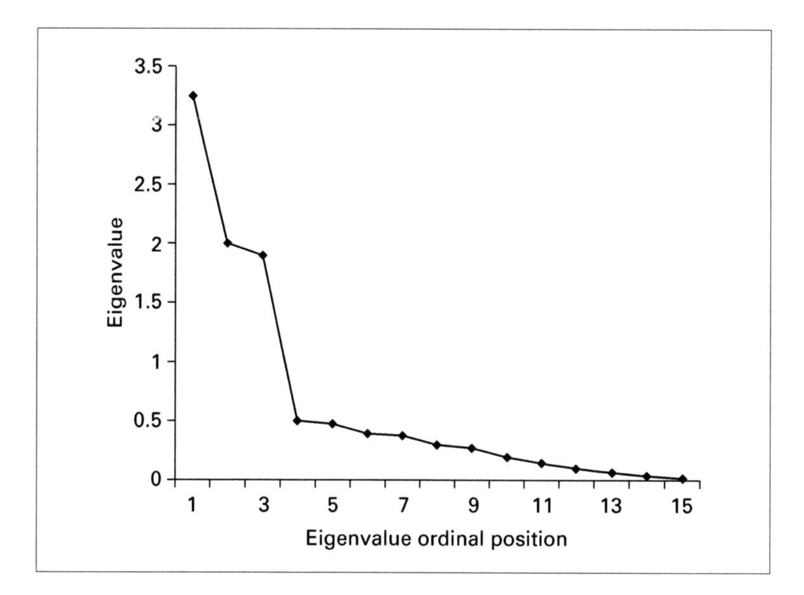

Figure 3.1. Hypothetical Example of a Scree Plot

or EFA is being conducted). We will address this issue in greater detail in Chapter 5.

The scree test has been criticized on the basis that it is a comparatively subjective procedure (e.g., Kaiser, 1970). For instance, there is no formal definition of what constitutes a major drop in a scree plot. Moreover, in some cases, the pattern of eigenvalues may be ambiguous in that no clear drop may be present at all, but rather the plot may show a series of steady incremental drops. These drawbacks notwithstanding, there is a plausible informal rationale underlying the procedure. Recall that eigenvalues represent the variance accounted for by a particular factor. The rationale for the scree plot is that, if there exists a specific number of *m* major common factors, then there will be *m* relatively large eigenvalues, and the remaining eigenvalues will be small in comparison. Those remaining small eigenvalues represent noise or the effects of random or small systematic influences. In an informal sense then, the scree test is directed at identifying the major common factors such that a model with too few factors should do worse in accounting for the data, and a model with too many factors should not do much better. Studies with simulated data have suggested that, when strong common factors are present in the data, the procedure works reasonably well (Cattell & Vogelmann, 1977; Hakistan et al., 1982; Tucker et al., 1969).

Parallel Analysis

Parallel analysis is a procedure that attempts to provide a more objective criterion than the scree plot and a less arbitrary criterion than the eigenvalues-greater-than-one rule for whether a particular factor constitutes more than noise in the data. Parallel analysis is based on the comparison of eigenvalues obtained from sample data with eigenvalues obtained from completely random data. Suppose that a battery of measured variables in a given sample depends on *m* major common factors. Parallel analysis is based on the notion that the *m* largest sample eigenvalues from the reduced correlation matrix should be substantially larger than the *m* largest eigenvalues from a corresponding set of random data (a random data set with the same sample size and number of measured variables).

Parallel analysis involves calculating the eigenvalues that would be expected from random data for a reduced correlation matrix with the same number of measured variables and based on the same sample size. These eigenvalues from random data are then compared with their corresponding eigenvalues from the real data (i.e., the first eigenvalue from the real data set is compared with the first eigenvalue from random data, the second eigenvalue from real data is compared with the second eigenvalue from random data, and so on). The appropriate number of common factors is the number of eigenvalues from the real data that are larger than their corresponding eigenvalues from random data.

A variety of procedures have been proposed and examined for generating the expected eigenvalues from random data (Allen & Hubbard, 1986; Buja & Eyuboglu, 1992; Cota, Longman, Holden, & Fekken, 1993; Cota, Longman, Holden, Fekken, & Xinaris, 1993; Glorfeld, 1995; Horn, 1965; Humphreys & Ilgren, 1969; Humphreys & Montanelli, 1975; Lautenschlager, 1989; Lautenschlager, Lance, & Flaherty, 1989; Montanelli & Humphreys, 1976; Timmerman & Lorenzo-Seva, 2011). Some approaches use regression equations to calculate the expected mean eigenvalues from random data for a given sample size and number of measured variables. Other approaches use simulation procedures to generate multiple random data sets of a given sample size and number of measured variables and then directly calculate the mean eigenvalues from these data sets. Additionally, some advocates of parallel analysis have suggested that, rather than using the mean eigenvalues from random data as the points of comparison, it might be more appropriate to use the somewhat more stringent criterion of the ninety-fifth percentile of the distribution of each eigenvalue. Although some important aspects of the relative performance of different parallel analysis procedures are yet to be fully explored, it is worth noting that regression-based approaches were largely developed as an alternative to simulated data approaches because of the computational challenges of the later methods. These difficulties have largely been eliminated with advances in computer technology.

As is the case with the scree test, one issue a researcher must consider is whether to base the parallel analysis on eigenvalues from the reduced correlation matrix or eigenvalues from the (unreduced) sample correlation matrix. Parallel analysis procedures have been developed to generate expected eigenvalues from random data for either of these matrices. In the context of EFA, parallel analyses procedures examining the reduced correlation matrix seem most conceptually sensible. In the case of PCA, parallel analysis with the unreduced correlation matrix is most logical.

Like other purely mechanical rules, parallel analysis can be somewhat arbitrary in that a factor just meeting the criterion might be retained, whereas a factor just below the criterion is ignored, despite the fact that the difference in the variances accounted for by the two factors might be trivial. Moreover, satisfying the criterion that the factor must account for more variance than would be expected by random data does not in any way guarantee that the factor will be readily interpretable or that inclusion of this factor will substantially improve the performance of the model over a model with one less factor. Indeed, one might argue that the parallel analysis criterion that a factor must simply outperform what would be expected from random data is a comparatively lenient standard by which to judge what constitutes a major common factor. As such, one might argue on purely conceptual grounds that parallel analysis as a general approach should perhaps be judged as a procedure for establishing the upper boundary of the number of common factors that should be considered for inclusion in the model. Moreover, traditional parallel analysis approaches based on the reduced correlation matrix using squared-multiple correlations in the diagonal (Humphreys & Ilgren, 1969; Humphreys & Montanelli, 1975) have been found to sometimes overfactor (e.g., see Buja & Eyuboglu, 1992; Timmerman & Lorenzo-Seva, 2011). More recently, a parallel analysis procedure based on minimum rank factor analysis has been found to be less susceptible to overfactoring in some contexts (Timmerman & Lorenzo-Seva, 2011).

These limitations notwithstanding, parallel analysis procedures have been found to perform reasonably well in simulated data sets (Allen & Hubbard, 1986; Humphreys & Montanelli, 1975; Lautenschlager, 1989; Longman et al., 1989; Timmerman & Lorenzo-Seva, 2011; Zwick & Velicer, 1986), although it should be noted that there are many conditions under which the performance of these procedures have yet to be evaluated. Additionally, although standard statistical programs do not include parallel analysis options in their factor analysis procedures, simple parallel analysis macros have been written for use with programs such as SPSS and SAS (O'Connor, 2000; see chapter 5 for an illustration). Standalone programs that are capable of implementing more recently developed parallel analysis procedures have also been developed (Lorenzo-Seva & Ferrando, 2006).

The Likelihood Ratio Test Statistic

When ML factor analysis is used, another potential approach to determining the number of factors is to examine the goodness of fit of the model. As noted earlier in the chapter, the likelihood ratio test statistic is one index of model fit provided by nearly all standard factor analysis programs that include ML model fitting procedures. Using the likelihood ratio test statistic, one can attempt to determine the appropriate number of factors in one of two ways. One approach is to begin by specifying the simplest factor-analytic model, a one-factor model, and then examine if the test statistic is significant. If it is not significant, the researcher fails to reject the model and, thus, a one-factor model would be retained. If the test is significant, the one-factor model is rejected and the researcher then fits a two-factor model. The process of specifying models of increasing complexity continues until a nonsignificant test statistic is obtained.

Although this approach has a clearly developed rationale and provides a very objective criterion for determining the number of factors, as we have already noted, the likelihood ratio test has significant drawbacks as a fit index. These limitations make it problematic as a basis for determining the number of factors. For example, as already noted, the criterion of perfect fit is not a realistic standard against which to judge a model. Thus, rejecting a model because it lacks perfect fit does not really indicate that additional major common factors exist. Furthermore, the sensitivity of the likelihood ratio fit index to sample size makes it likely that models will be substantially overfactored in large sample sizes (where even very small discrepancies between the model and the data might lead to rejection of the model) and underfactored in small sample sizes (where even comparatively large discrepancies between the model and data might fail to be detected). Additionally, the approach can lead to fairly arbitrary decisions. For instance, if treating a *p*-level of .05 as the cutoff value, a model producing a test statistic with a *p* value of .04 would be rejected, whereas a model with one more factor that produced a *p* value of .06 would be accepted. Such a decision seems difficult to defend, given that the addition of the factor resulted in only a trivial difference in improvement of the model.

A variant of the approach just described involves using the likelihood ratio test statistic to conduct "difference tests" between models of increasing complexity. In this approach, the researcher begins with a one-factor model and obtains a likelihood ratio test statistic for this model. The researcher then fits the two-factor model and computes the test statistic for this model. Because an EFA model with fewer factors is a special case of a model with more factors (i.e., the first model is nested within the second), it is possible to conduct a statistical test to see if the addition of the factor leads to a statistically significant improvement in fit. This is done by computing the likelihood ratio χ^2 difference test between models where:

$$\chi^2 = \chi^2 \text{ (of model with fewer factors)} - \chi^2 \text{ (of model with more factors)}$$

$$df = df \text{ (of model with fewer factors)} - df \text{ (of model with more factors)}$$

The difference between the χ^2 values of the two models also follows a χ^2 distribution with a df equivalent to the difference between the df of the two models. Thus, this new χ^2 can be examined to determine if the test is significant. If the test is not significant, the model with one less factor is retained. If it is significant, the model with one less factor is rejected in favor of the more complex model. This more complex model is the new comparison point against which a model with one additional factor is compared. The process terminates when a nonsignificant test is obtained. Thus, factors are only included if they provide a significant improvement over a model with one fewer factor.

This second approach to using the likelihood ratio test statistic has some advantages over the first approach in that models are not required to be perfect to be retained. A model must simply do better than a model with one fewer factor and not significantly worse than a model with one more factor. That being said, the sensitivity of the index to sample size still poses a problem in that large sample sizes may produce statistically significant improvements in fit that in practical or conceptual terms are comparatively trivial. Conversely, the test may fail to detect sizeable improvements in fit when sample sizes are small.

Descriptive Indices of Model Fit

As noted earlier, the fit of any ML factor analysis model can also be evaluated using descriptive fit indices commonly employed in CFA and SEM. Browne and Cudeck (1992) recommended using descriptive fit indices as a method of determining the appropriate number of factors (see also Fabrigar, Wegener, MacCallum, & Strahan, 1999). Of the specific indices discussed by Browne and Cudeck (1992) for this purpose, probably the most widely used is RMSEA. RMSEA can be used to determine the appropriate number of factors by specifying a series of models of increasing complexity. The researcher begins with a one-factor model and then specifies a sequence of models, each of which has one additional factor. The upper limit of model complexity specified in the sequence is usually the point at which additional common factors would cease to be conceptually useful or for which the model would approach a point at which it would have zero degrees of freedom. The appropriate number of factors is then determined by examining the RMSEA values for the sequence of models. The appropriate number of factors is that model in the sequence of models that ideally: (1) fits the data well, (2) fits the data better (has a substantially lower RMSEA) than a model with one less factor, and (3) fits the data about as well as (has an RMSEA that is not substantially higher) a model with one more factor.

Several practical issues arise in judging these three criteria. First, how good must the fit of the model be in order to be considered as a potential candidate for having the appropriate number of common factors? Based on the guidelines for interpreting RMSEA discussed earlier in the chapter, a reasonable standard would be that a model should have an RMSEA equal to or lower than .080 (i.e., a model with acceptable fit), and ideally should have an RMSEA of .050 or lower (i.e., a model with close fit).

A second issue that must be addressed in implementing the aforementioned criteria is what constitutes a substantially better or worse RMSEA value? That is, how big must the difference in RMSEA be between two models to be considered meaningful? Using the existing guidelines for interpreting RMSEA as a reference point, it is worth noting that the various categories of fit specified in these guidelines range in width from 0.02 to 0.05. Thus, consistent with these guidelines, any difference of 0.020 or greater can generally be considered as a substantial difference in fit. Differences of 0.010 to 0.019 might be considered as marginally meaningful differences in fit, but differences less than 0.010 rarely constitute meaningful improvements in model fit. Additionally, although it is the magnitude of difference in RMSEA that is critical for distinguishing between models using this approach, researchers should not ignore the confidence intervals associated with RMSEA values. When confidence intervals are broad (e.g., when two models with even substantial differences in the magnitude of their RMSEA values are still contained within the confidence intervals of each other), researchers should recognize that insufficient precision exists in estimates of fit to confidently make comparisons between models or to evaluate the fit of any model in an absolute sense. Thus, any conclusions regarding model fit reached in such a context should be treated with caution.

Yet another practical challenge that can sometimes arise when using this approach is situations in which the absolute standard of acceptable/good fit for a model conflicts with the relative comparison of a substantial improvement in fit between two models. In some contexts, major improvements in the fit of the model with the addition of another factor might stop at a point where the model still fits the data rather poorly. Thus, adding another factor might not improve the model much, but failing to add more factors might leave the researcher with a model that has poor fit. Such an occurrence might suggest that there is no small number of common factors sufficiently strong to effectively account for the data. Thus, there could be a large set of common factors of moderate to low strength such that no single factor contributes a great deal on its own to explaining the data, but collectively the factors have substantial explanatory power.

To date, there are no published studies examining this approach of using RMSEA as a means of determining the number of common factors in EFA. However, the approach does have a very compelling rationale. Additionally, RMSEA has been found to be a comparatively useful index in detecting model misspecification in SEM (e.g., see Hu & Bentler, 1998) and preliminary tests of its utility in determining the appropriate number of common factors in the context of EFA have been encouraging (Thompson, 2004). It should also be noted that this general approach of using descriptive model fit indices to determine the appropriate number of factors need not be restricted to RMSEA. In principle, any descriptive model fit index that is used in structural equation modeling can be computed for ML EFA and used as the basis for determining the appropriate number of factors (e.g., see Fabrigar et al., 1999; Lorenzo-Seva, Timmerman, & Kiers, 2011).

Stability of Solutions

For contexts in which a researcher has multiple data sets available to test a model or a very large data set that could be split in half, the stability of a factor solution can be a useful criterion for determining the appropriate number of factors (see Cattell, 1978). For example, imagine that a researcher has a situation in which the aforementioned factor number procedures produce somewhat ambiguous results such that there are two or three models that are viable representations of the data. The researcher can fit these competing models to each of the data sets and examine the final rotated solutions for the models to see if one solution is more stable across data sets. The model that produces the most replicable pattern of parameter estimates over the data sets is judged to be the model with the appropriate number of factors.

Interpretability of Solutions

A final consideration that is crucial to evaluating the appropriate number of factors is the interpretability of the common factors. Ultimately, a factor analysis model is only useful if it provides a conceptually sensible representation of the data. Thus, in order for a model to be judged to have the appropriate number of common factors, it is important that the final rotated solution for each factor be readily interpretable. For cases in which statistical approaches to determining the number of factors provide ambiguous or conflicting results, the interpretability and theoretical utility of the competing models often will be the deciding criterion in selecting among the models.

General guidelines for interpreting factor analysis models are difficult because any model must be interpreted in light of the nature of the particular measured variables being examined. Thus, a researcher's substantive knowledge of the area of interest will be critical to interpreting the resulting solution. That being said, some general principles of factor interpretation can be advanced.

First, when naming a common factor (i.e., interpreting a common factor as representing a particular construct), it is important to recognize that all measured variables substantially loading on that common factor must be plausibly judged as reflecting the hypothesized construct. That is, there must be a common theme to all measured variables loading on the same factor. Equally important, all measured variables not loading on that common factor must be plausibly judged as unlikely to reflect the hypothesized construct. That is, no measured variables failing to load on the factor should also share this theme with the measured variables that do load on the common factor.

A second consideration to keep in mind is that measured variables loading on more than one common factor are not in and of themselves problematic. Such measured variables can often be quite interpretable if the measured variable can be plausibly interpreted to be influenced by more than one construct. For some measured variables, this is not only reasonable, but to be expected. For example, in the context of mental abilities, one might expect that a test consisting of mathematical word problems should be influenced by both a verbal ability common factor and a mathematical ability common factor because the test requires both skills to successfully complete it. Thus, measured variables with substantial loadings on multiple factors are only problematic if it is difficult to conceptualize that the constructs postulated to be represented by the common factors could all be exerting influence on the measured variable.

Third, common factors that have no measured variables with substantial factor loadings or only a single measured variable with a substantial loading are generally difficult to interpret and often reflect overfactoring. In the first case, it is difficult to interpret a common factor as a major common factor if it does not strongly influence any measured variables. Similarly, in the second case, because common factors by definition should influence more than one measured variable, a common factor that only strongly influences one measured variable is difficult to define as a major common factor.

Fourth, for situations in which it is difficult to arrive at a single common theme to a set of measured variables that load on a common factor, this can sometimes be indicative of underfactoring. Often when too few common factors have been specified in the model, two or more factors will be collapsed onto the same factor, making it difficult to identify a single unifying theme among the measured variables (see Comrey, 1978; Wood, Tataryn, & Gorsuch, 1996).

Summary

A wide array of procedures is available to researchers for determining the appropriate number of factors.[3] Some of these approaches are highly flawed and are probably best avoided (i.e., the eigenvalues-greater-than-1 rule, the likelihood ratio test statistic). Other rules have more utility and can offer the researcher useful information (i.e., the scree test, parallel analysis, RMSEA, factor stability, and factor interpretability). It is important to recognize that even these better performing procedures are not infallible. Each procedure approaches the question of determining the number of factors in a somewhat different manner. Thus, one procedure should never be used in isolation. Rather the decision on the number of factors to retain should be based on the totality of the evidence presented by these procedures.

Rotating Factor Analysis Solutions

As we noted in our discussion of model fitting procedures, for any EFA model (or PCA model) with two or more common factors (or principal components), there will exist an infinite number of equally fitting solutions for the parameter estimates of the model. That is, there will be an infinite number of alternative orientations of the common factors (or principal components) in multidimensional space. The implication of this fact is that there will not be a unique solution for the parameter estimates of any model with more than one factor. Thus, the researcher must select one solution from the infinite number of equally best-fitting solutions. This challenge, which has been termed *rotational indeterminacy*, is perhaps best illustrated by presenting the common factor model in its geometric representation. We illustrate the model in the context of a two-factor model (i.e., a 2-dimensional spatial representation), but the logic can be readily extended to three or more factors.

Any factor analysis model can be conceptualized as a spatial representation of the measured variables in which the common factors constitute the dimensions or axes of the space and the factor loadings reflect the coordinates of the measured variables in this space. Figure 3.2 illustrates this idea in the context of a two-factor model with 12 measured variables. In the geometrical representation, the distance between measured variables in this spatial representation represents the strength and direction of correlations among measured variables. Measured variables that are very close to one another have very strong positive correlations. Measured variables that are maximally distant from one another have strong negative correlations. The fit of the model is reflected in the extent to which the physical distance in space between measured variables can accurately reconstruct the observed correlations among measured variables. If the mathematical function for converting physical distance into a predicted correlation between measured variables comes very close to the actual correlations among the measured variables in the sample, the model fits well. If it does not, it suggests that a two-dimensional spatial representation is inadequate and that we might for example, have to represent

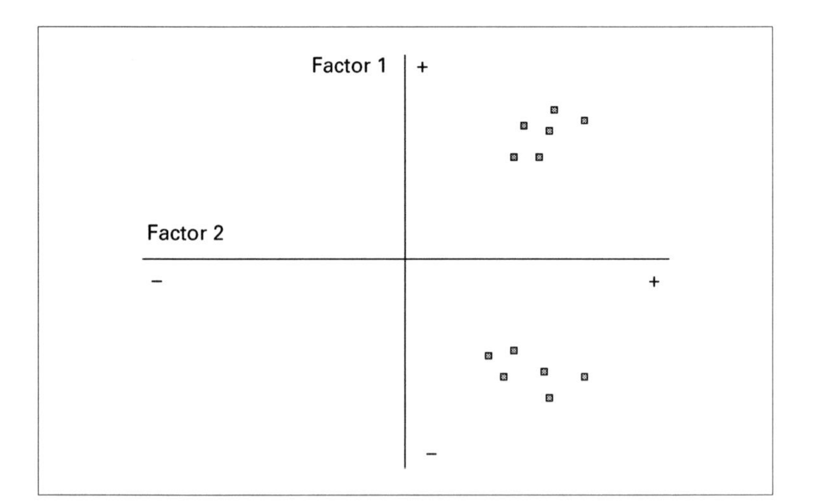

Figure 3.2. Geometrical Representation of a Two-Factor Model

our data using a more complex form such as in three-dimensional space (i.e., a three-factor model).

A critical feature of the representation in figure 3.2 is that the ability of this spatial representation to account for our correlations is not dependent on the orientation of the two dimensions. That is, we could rotate these dimensions to any other orientation and these orientations would do equally well at explaining the correlations (see the dashed lines in figure 3.3). A re-orientation of the two dimensions would, of course, require a recalculation of a new set of coordinates (i.e., factor loadings) to reflect the new orientation of the axes (dimensions), but this new set of coordinates would be just as good as the first set in accounting for the data. That is, the re-orientation would not change the fundamental distances in two-dimensional space that represent and estimate the correlations among measured variables. Thus, there will be an infinite number of equally best-fitting solutions for the two-factor model depicted in figure 3.2 and 3.3.

One of the challenges in the EFA literature is to decide which orientation of the common factors (dimensions or axes) is most useful. That is, how do we select a single solution among this infinite set of equally best-fitting solutions? A temporary solution to this problem was discussed in our section on model fitting procedures. When the

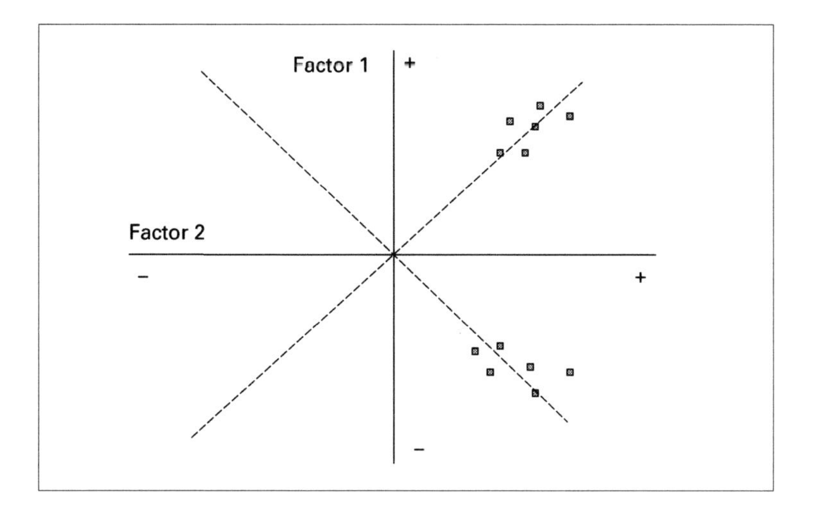

Figure 3.3. Geometrical Representation of a Two-Factor Model with a Rotation of Axes

model is initially fit to the data, a set of identification conditions are specified such that the solution must satisfy certain mathematical criteria. These mathematical criteria will be satisfied by one and only one solution (i.e., orientation of dimensions) among the infinite set of best solutions. However, it is important to recognize that the specific solution selected as a function of these identification conditions is not chosen on the basis of its conceptual plausibility or ease of interpretation but, rather, by its computational convenience. The question then arises about whether computational ease should be the criterion by which we judge the best solution. The answer provided by most researchers using EFA is no. Rather, most methodologists argue that the researcher should select the solution that is most theoretically plausible, readily interpretable, and likely to replicate across studies.

Simple Structure

Of course, this response raises a new question about what sort of solution should be most likely to satisfy these criteria? To some degree, the answer to this question will be determined by the nature of the measured variables being investigated. Hence, there is no single correct answer. However, there may be general properties of a factor analysis solution that tend to make it more readily interpretable and plausible in most contexts.

Of the properties that have been proposed, far and away the most widely accepted is the concept of *simple structure* originally advanced by Thurstone (1947). Thurstone suggested five general guidelines for a solution with simple structure (see also Browne, 2001):

1. Each row (i.e., each measured variable) in the factor loading matrix should contain at least one zero (or a value very close to 0).
2. Each column (i.e., each common factor) should contain at least m zero or near-zero values (where m represents the number of common factors).
3. Every pair of columns of the factor loading matrix (i.e., every pair of common factors) should have several rows with a zero or near-zero value in one column but not in the other.
4. For situations in which m is equal to or greater than four, every pair of columns of the factor loading matrix should

have several rows with zero or near-zero values in both columns.
5. Every pair of columns of the factor loading matrix should have only a few rows with non-zero loadings in both columns.

When taken as a whole, these guidelines imply three general properties of factor solutions with simple structure. First, in a simple structure solution, each common factor is represented by a distinct subset of measured variables with large factor loadings and for which the remaining measured variables have small factor loadings. Second, the subsets of measured variables defining the different common factors should not substantially overlap with one another. Finally, each measured variable should be influenced by only a subset of common factors. An important point to recognize, which is often misunderstood, is that the Thurstone guidelines for simple structure do not require a measured variable to load substantially on only one common factor, particularly when dealing with factor analytic models with three or more factors. The guidelines merely indicate that a measured variable should not load on all the common factors. As we discussed earlier, there are a number of contexts where one might conceptually expect a measured variable to be influenced by more than one common factor. A solution in which each measured variable substantially loads on only one common factor is referred to as an *independent cluster solution* and constitutes a special case (but not the only instance) of simple structure.

Thurstone argued that, for any given set of equally best-fitting solutions, in general, the solution that best exemplified these properties would be the solution that was most easily interpreted, most conceptually meaningful, and most likely to replicate. Thus, Thurstone recommended that researchers rotate the axes of their initial factor analysis solutions to maximize simple structure and base their interpretations of their analyses on this rotated solution. In the early days of the factor analysis literature, this was accomplished by hand. That is, a researcher examined the visual display of the solution and then subjectively judged what change in orientation would best exemplify simple structure. After having determined the angle of rotation, new factor loadings were then computed by hand to reflect the changed orientation of the common factors.

Consider the solution depicted in figure 3.3. The orientation of the common factors in this initial solution (depicted by the solid-line axes) does not exemplify simple structure. The measured variables loading in the upper-right region of the solution have substantial positive loadings on both factors. The measured variables located in the lower-right region of the solution have substantial positive loadings on factor 2 and substantial negative factor loadings on factor 1. Thus, neither common factor is defined by a distinct subset of measured variables, and all of the measured variables are influenced by all of the common factors. Such a solution would be very difficult to interpret. However, if the factors are rotated (see the dashed axes in figure 3.3) such that one factor axis passes through the measured variables in the upper-right region of the solution and the second factor axis passes through the measured variables in lower-right region of the solution, the simple structure of the solution is greatly enhanced. The upper-right grouping of measured variables now has substantial loadings on factor 1 and small loadings on factor 2. The lower-right grouping of measured variables now has small loadings on factor 1 and large loadings on factor 2.

Unfortunately, this simple visual inspection and hand calculation approach to rotation presented several problems. First, it was very time consuming, particularly as the number of measured variables increased. Second, it often involved a substantial component of subjectivity. The solution depicted in figure 3.3 provides a comparatively clear pattern of results, and it is easy to see what change in orientation will improve simple structure. As solutions become more complex with less well-defined groupings of measured variables, it becomes more difficult to readily discern the best orientation of the axes. Finally, the approach becomes increasingly unwieldy at higher levels of dimensionality. Representing data in two dimensions is quite simple. Representing data in three dimensions is more complex, but still manageable. However, factor models of four or more dimensions become extremely difficult to visually represent and then judge how changes in the orientations will alter simple structure.

Orthogonal Analytic Rotation

Because of the difficulties associated with rotations using visual displays, by the 1950s researchers began to explore automated methods of rotating solutions. This automated approach, referred to as *analytic rotation*, involves defining some mathematical property of the factor loading matrix that is thought to be related to the extent to which the solution will manifest simple structure. A given rotation procedure then calculates which of the infinite number of equally best-fitting solutions maximizes that mathematical property. The specific mathematical function that a given rotation procedure attempts to maximize is referred to as the *simplicity function*. Because signs of factor loadings depend on arbitrary choices of scoring direction of factor variables, it is customary for simplicity criteria to be based on squared factor loadings. Importantly, this approach to rotation is completely objective in that it does not require the user to make subjective judgments about the ideal orientations of factors, and the approach can be computerized so that it can be performed quite rapidly.

The first simplicity criterion to be proposed was quartimax rotation, which was independently developed by several researchers (Carroll, 1953; Saunders, 1953; Neuhaus & Wrigley, 1954; Ferguson, 1954). This rotation assumed orthogonal common factors. Unfortunately, it was found to sometimes produce solutions with a single general factor (i.e., a common factor strongly influencing all measured variables) and other common factors in which all measured variables had small loadings.

In response to this problem, Kaiser (1958) proposed *varimax rotation*. Varimax rotation is based on the premise that, as the simple structure of each factor improves, the squared loadings on those factors become more variable (i.e., some loadings are very large and the rest very small). Thus, the simplicity criterion of varimax is the sum of the variances of the squared factor loadings across all the common factors in the model. Varimax rotation solves for a solution among the equally best-fitting solutions with maximum variability in the factor loadings. This form of varimax rotation is often referred to as *raw varimax* rotation because it is applied to the raw factor loadings prior to processing. Raw varimax was found to function reasonably well in many contexts, but did tend to produce solutions with poor simple structure for measured variables with small communalities. Kaiser, therefore, proposed standardizing rows of the factor loading matrix by dividing factor loadings by the square roots of communalities before rotation and then multiplying the rotated factor loadings by the same communalities after rotation. This procedure is referred to as *normal varimax* or *varimax with row standardization.* It is the version of varimax rotation most commonly implemented by standard statistical programs and it has been found to usually function well when the procedure's assumption of orthogonal factors is met. Normal varimax remains probably the most widely used method of rotation.

Oblique Analytic Rotation

Recall from our discussion of model fitting procedures that, when these fitting procedures produce their initial estimates of the model parameters, they do so under the assumption that all common factors in the model are uncorrelated. Orthogonal rotations, such as varimax rotations, transform the initial factor loading matrix, retaining this assumption of uncorrelated factors. That is, they attempt to find the single solution among the array of equally best-fitting orthogonal solutions that demonstrates the best simple structure, rather than to find the single solution with the best simple structure among all possible equally best-fitting solutions. However, very early on in the rotation literature, it was recognized that such a restriction was often unrealistic. For example, in Thurstone's writings on simple structure (Thurstone, 1935, 1947), he placed an emphasis on *oblique rotation* (i.e., rotations permitting common factors to be correlated) and argued that correlated factors were often a more plausible representation of the data (see Browne, 2001). Indeed, in many of the contexts in which psychologists and other social scientists have used EFA (e.g., mental abilities, the structure of attitudes), there is a substantial theoretical basis to expect correlations among common factors. In fact, these correlations among the common factors may themselves be a focus of substantive interest and the pattern of correlations among common factors may provide useful insights into the nature of the common factors.

Despite these facts, reviews of the applied factor analytic literature indicate that orthogonal rotations (and varimax in particular) have been the most prevalent approach to rotation (e.g., Fabrigar et al., 1999; Ford et al., 1986). Why is this? One reason may be that, although the problem of finding a satisfactory orthogonal analytic rotation was solved comparatively quickly with the advent of varimax rotation, the task of developing a satisfactory method of oblique rotation proved more challenging (Browne, 2001). Thus, varimax became well established within the factor analytic literature and it was only later that viable oblique rotations were available to researchers. It is likely that many subsequent researchers adopted varimax rotation simply because it had been used in earlier studies.

Second, a number of misconceptions exist about the distinction between oblique and orthogonal rotation. For example, some researchers incorrectly believe that oblique rotations require or cause common factors to be correlated. Rather, oblique rotations allow for or permit common factors to be correlated. Oblique rotation is simply a more general form of rotation that allows for both uncorrelated and correlated common factors. The solution that best manifests simple structure (as defined by the rotation's simplicity criterion) is selected among all possible equally best fitting orthogonal and oblique solutions. If the factors are uncorrelated and simple structure is maximized with an orthogonal solution, successful oblique rotations will estimate the common factor correlations to be near zero and produce a solution very close to a varimax rotation (see Harman, 1976). However, if better simple structure can be obtained with a solution involving correlated common factors, the oblique rotation will identify such a solution.

Another misconception regarding oblique rotation is that because orthogonal common factor models are "simpler" than oblique models, orthogonal rotations will produce better simple structure. In point of fact, the opposite is true. When the common factors underlying the data are correlated, oblique rotations will usually produce better simple structure than orthogonal rotations. This point can be illustrated in at least two ways. First, a common factor model can be thought of as a model in which a set of latent variables (common factors) are used to predict a set of measured variables. In the case of an oblique model, the estimates of the influence of a given common factor's influence on the measured variables (i.e., the factor loadings) is an estimate of that common factor's influence holding constant or controlling for the influence of the other common factors in the model. Thus, these factor loadings can be thought of as similar to partial regression coefficients in multiple regression. As such, an oblique rotation eliminates spurious effects of the common factors on the measured variables (i.e., a nonzero factor loading that emerges, not because that factor influences the measured variable, but simply because the factor is correlated with another factor that does influence the measured variable). Elimination of spurious effects will tend to enhance the simple structure of solutions.

A second way to illustrate this point is to consider the geometrical representation of the common factor model. When represented in this way, the fundamental difference between orthogonal and oblique rotations is that orthogonal rotations require the angle between axes to remain at 90°. In contrast, oblique rotations permit the angle to be less than 90° (with correlations among factors increasing as the angle is reduced) if better simple structure can be obtained by doing so.

In figure 3.3, the restriction of orthogonal factors does not pose a problem, because the two clusters of measured variables are located approximately 90° from one another. Thus, we can rotate the orthogonal axes to an orientation that produces good simple structure. However, consider the situation depicted in figure 3.4. In this case, the clusters of measured variables are now located at less than 90°. In other words, scores on the first cluster of measured variables will tend to be correlated with scores on the second cluster of measured variables (i.e., the underlying common factors are correlated). In this situation, no orientation of the axes with a

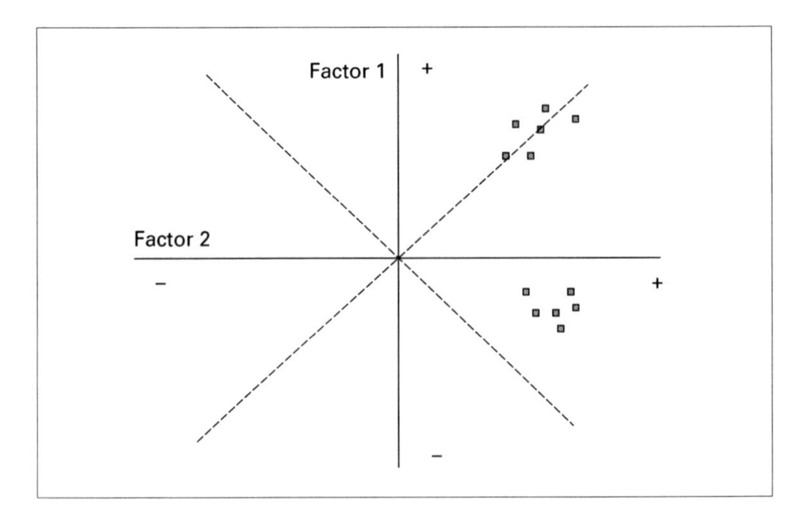

Figure 3.4. Geometrical Representation of a Two-Factor Model with an Orthogonal Rotation of Axes

90° angle can produce especially good simple structure. We could rotate our dimensions so the first factor passes through the cluster of measured variables in the upper-right region, but this would result in the second factor not passing through the cluster of measured variables located in the lower-right region. Alternatively, we could rotate so that the second factor passes through the lower-right region of measured variables, but this would lead to poor simple structure for the measured variables in the upper-right region.

In contrast, an oblique rotation allows us to reduce the angle thereby achieving good simple structure (see figure 3.5). The extent to which the angle must be reduced to achieve simple structure allows us to calculate the correlation between the common factors.

A final misconception that sometimes exists among researchers regarding rotation is that if a researcher believes that the factors are or wants the factors to be uncorrelated, orthogonal rotation can, in some way, accomplish this objective. If two underlying common factors are in reality correlated with one another, using a rotation that assumes them to be uncorrelated does not change this fact. Model assumptions do not alter data. Once again, this can be readily illustrated using the geometric representation of the data. The spatial relations between the two clusters of measured variables depicted in figures 3.4 and 3.5 are what they are.

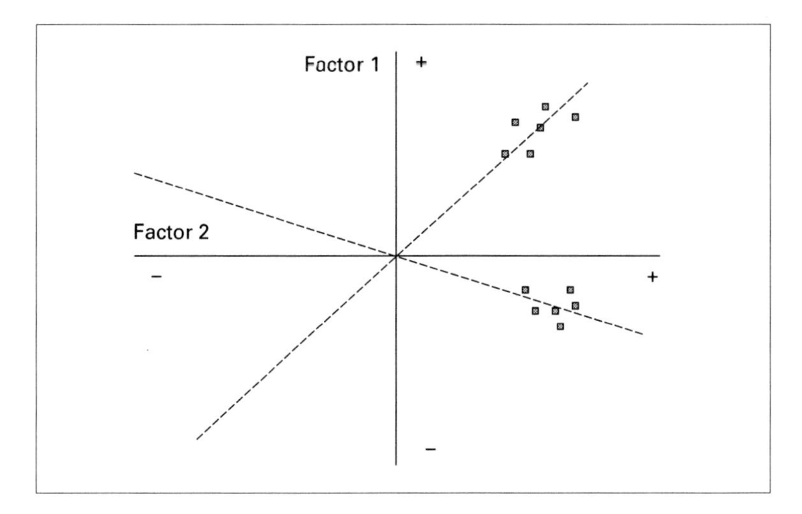

Figure 3.5. Geometrical Representation of a Two-Factor Model with an Oblique Rotation of Axes

Whether we conduct a rotation that restricts the angle to be 90° or that permits a smaller angle between the dimensions in no way alters these relations.

Given these facts, oblique rotation is generally a more sensible approach than orthogonal rotation. Oblique rotations will often be a more realistic representation of the data, will provide a solution that allows for easier interpretation, and will give the researcher additional information not available in orthogonal rotations (i.e., the correlations among common factors). Fortunately, viable oblique rotations are now available.

One of the first oblique rotation methods that was found to function adequately in many contexts and received widespread usage was *Promax rotation* (Hendrickson & White, 1964). Promax rotation begins by conducting an initial Varimax rotation and then conducts a mathematical transformation of this initial solution, by raising factor loadings to a power of two or greater, to arrive at an oblique solution. The specific power by which factor loadings is raised is defined by the kappa parameter (k) of the rotation, with higher values producing greater correlations among factors. A value of four is most commonly used and this is generally the default value used in most statistical programs. Promax rotation is available in both the SPSS and SAS programs.

Another oblique rotation that has been found to function adequately in many contexts and has been widely used is the *Harris-Kaiser rotation* or *Orthoblique rotation* (Harris & Kaiser, 1964). Like Promax, this method begins with an initial Varimax rotation. This solution is then re-scaled to arrive at an oblique solution. The Harris-Kaiser rotation also requires the user to specify a parameter, with different values of the parameter rescaling the solution with different goals. This parameter is often referred to as the HK Power or HKP. An HKP of 1 is equivalent to a Varimax rotation. An HKP of 0.5 is a general oblique rotation. An HKP of zero is an oblique rotation that seeks an independent cluster solution. In general, an HKP of 0.5 is probably the value that is most sensible to use in most contexts. Orthoblique rotation is available in the SAS program.

Both Promax and Orthoblique rotation are indirect methods of rotation in that they have an intermediate step in rotation. Direct quartimin rotation (Jenrich & Sampson, 1966) was the first successful oblique rotation to rotate directly to simple structure and is preferred by some methodologists over indirect methods of oblique rotation (e.g., see Browne, 2001). This rotation has been found to generally function well. It is a member of a family of rotations known as direct oblimin rotation, in which each member is defined by the value of a constant, known as delta, which specifies the level of obliqueness among factors. Direct quartimin has a delta of 0, which indicates an equal weighting of correlated versus uncorrelated factors. It is the default value for delta in the direct oblimin rotation procedure in SPSS. Not all members of this family of rotations functions well, so researchers should generally use the direct quartimin criterion when using this procedure.

Interpretational Issues with Oblique and Orthogonal Rotations

When considering both orthogonal and oblique rotations of solutions, there are two facts that researchers should keep in mind. First, the scaling direction of common factors is arbitrary. Programs simply scale solutions on the basis of computational convenience rather than as a function of some fundamental conceptual property of the common factors. Thus, for any factor solution, it is permissible to reverse the signs of all factor loadings in a given column of the factor loading matrix. Such reversals (as long as they are applied to all elements in the column) do not affect the fit of the model or the communalities of the measured variables. Researchers are sometimes confused when they obtain a solution in which all measured variables load negatively on a common factor even though all of the measured variables are keyed in a positive direction such that higher scores indicate more of the hypothesized construct. Such a result merely indicates that the program scaled the common factor in the reverse direction (lower numbers indicating more of the underlying construct) and it would be entirely reasonable for the researcher to reverse the signs of all the factor loadings to arrive at a solution that is more intuitively scaled.

A second issue to keep in mind when interpreting both orthogonal and oblique solutions is that the ordering of factors following rotation is generally arbitrary. That is, although the ordering of factors in the initial unrotated solution is meaningful in the sense that each successive factor explains less variance than the prior factor, this property need not be true after rotation. Rotation will not alter the fit of the model, the overall amount of variance accounted for by the common factors, or the communalities of the measured variables. However, it will redistribute the variance accounted for by the factors across the rotated factors such that the first factor may no longer explain more variance than the second factor and so on. Likewise, although the communality of a given measured variable will not change after rotation, the amount of common variance for that measured variable explained by each of the common factors might be redistributed. Given these facts, it is entirely reasonable for researchers to re-order the rotated common factors for ease of interpretation. Indeed, researchers sometimes mistakenly conclude that they have not replicated a prior analysis because the order of the common factors is different. Generally, a proper replication in EFA only requires that the same common factors have been produced with comparable factor loadings and does not require the same ordering of these factors.

When interpreting rotated solutions, it is also useful to keep in mind some differences between oblique and orthogonal rotated solutions. First, in the context of orthogonal solutions, the factor loadings can be interpreted as correlations between the common factors and the measured variables. Thus, they are bounded by −1.00 and 1.00 and the proportion of variance in a measured variable accounted for by a given common factor can be calculated by squaring the factor loading. In the context of oblique rotations, the factor loadings no longer represent correlations between common factors and measured variables. Instead, the factor loadings are comparable to standardized partial regression coefficients. Thus, they reflect the standardized units of increase in a measured variable given one standardized unit of increase in the common factor, controlling for (partialing out) the effects of all other common factors in the model. Because oblique rotation factor loadings are similar to standardized partial regression coefficients, they are not bounded by −1.00 and 1.00 (although they will rarely go much beyond these values), and squaring them will not produce the proportion of variance in the measured variable accounted for by the common factor.

A second difference to keep in mind is that when conducting an orthogonal rotation, the procedure will produce a single rotated factor loading matrix. However, following oblique rotations, it is customary for programs to report three matrices: the pattern matrix, structure matrix, and the factor correlation matrix. The pattern matrix corresponds to the rotated factor loading matrix (Λ). Thus, it represents the influence of each common factor on the measured variables controlling for the effects of the other common factors in the model. The structure matrix represents the zero-order correlations between the common factors and the measured variables (i.e., the correlations between each common factor and the measured variables without controlling for the influence of other common factors in the model). Thus, the structure matrix does not contain the parameter estimates for Λ, nor have these values been specifically rotated for simple structure.

There has sometimes been confusion among researchers regarding which matrix should be the primary basis for interpretation, with some researchers actually choosing to focus on the structure matrix rather than the pattern matrix. Although it is certainly reasonable for a researcher to examine both matrices, the logic of oblique rotation suggests that the pattern matrix should be the primary basis for interpreting factors. It is these values that are the actual parameter estimates of the factor loading matrix in the common factor model. Attempting to interpret the results of a model using indices that do not directly correspond to the actual parameters of the model is difficult to justify on conceptual grounds. Moreover, choosing to focus on the structure matrix is inherently inconsistent with the primary goal of oblique rotation. The purpose of oblique rotation is to arrive at a solution with simple structure that takes into account correlations among factors and thus arrives at estimates of the unique effects of each common factor on the measured variables. The pattern matrix accomplishes this objective, whereas the structure matrix does not. Indeed, focusing on the structure matrix would be similar to a researcher conducting a multiple regression analysis to examine the unique predictive ability of each variable in a set of predictor variables, but then to ignore the partial regression coefficients and instead merely examine zero-order correlations of the predictor variables with the dependent variable.

The third matrix reported by oblique rotations is the factor correlation matrix. This matrix is the matrix of correlations among the common factors and corresponds to Φ in the common factor model. It can be interpreted much the same way as any matrix of correlations. However, it is useful to keep in mind that these are correlations among latent variables and thus reflect associations after having controlled for attenuation due to unique factors (and controlling for measurement error). As such, these correlations could, in theory, be expected to reach a value of 1.00 if two common factors were in fact entirely redundant.

A final point to consider in our discussion of rotation is that we have focused on rotations that attempt to maximize simple structure. This objective is certainly the most common criterion for rotation, and it is sensible in many contexts. Rotations such as direct quartimin tend to do well when the underlying structure of the data is a reasonably close approximation of simple structure. However, as the structure of data becomes more complex (i.e., many measured variables are substantially influenced by multiple factors), the performance of these rotations will erode and they can produce somewhat different results (Sass & Schmitt, 2010). Moreover, it is important to recognize that achieving simple structure will not always even be a sensible goal in some research contexts. For example, there are some cases in which we would not necessarily expect simple structure to be the most plausible or interpretable solution for our data (e.g., circumplex structure; see Fabrigar, Visser, & Browne, 1997). In such cases, one might not rotate a solution at all or would rotate it with some other goal in mind.

Of the various types of nonsimple structure rotations that are used, probably the two most common are target rotation (also called procrustes rotation) and confactor rotation. Target rotation is a quasi-confirmatory approach to rotation in which the researcher specifies which factor loadings are expected to be near zero and which are not. The specified pattern of zero and non-zero loadings might or might not reflect simple structure, depending on the theory guiding the specification process. The procedure then rotates to identify which of the equally best-fitting solutions comes closest to the specified pattern of zero and non-zero loadings. Indices of the extent to which the obtained solution matches the specified target can then be computed. Both orthogonal and oblique target rotation procedures have been developed. In confactor rotation, two simultaneous factor analyses are conducted on two independent samples, and then the resulting solutions in the two groups are rotated to be as close as possible to one another. There is no necessary requirement that the solutions that are closest to one another across the samples reflect simple structure.

Concluding Comments

We have now concluded our discussion of the requirements and decisions that researchers must consider when conducting an EFA. As much as any other statistical procedure commonly used in the social sciences, EFA requires informed judgment on the part of the researcher. At each phase of the analysis, researchers have an array of issues to consider and procedures from which to choose. These choices are not arbitrary in that they often rest on different assumptions, provide different information, and can lead to different substantive conclusions. Moreover, in many cases, not all these choices will be conceptually sensible or methodologically defensible. Thus, when conducting EFA, it is essential for researchers to carefully consider their choices at each phase of the process. The same is equally true when consuming the research of others. An EFA is only as good as the data on which it is based and the procedures by which it was conducted. Readers should carefully attend to the choices made by researchers in conducting their EFAs when evaluating the credibility of results reported.

Notes

1. Note that Λ^T is not an additional set of unknowns because it merely involves an alternative expression of the values of Λ. Thus, if the values of Λ are known, the values of Λ^T are also known.
2. In recent years, there has been work conducted on developing versions of IPA and NIPA that permit calculation of standard errors and significance tests (see Browne et al., 2010). However, these versions of IPA and NIPA also assume multivariate normality and the extent to which they are robust to violations of this assumption has yet to be fully evaluated.
3. One often advocated approach to determining the number of factors that we have not discussed is the minimum average partial (MAP) test proposed by Velicer (1976). Because this approach was developed for use with PCA rather than EFA, we have not reviewed it here. Recent research exploring its performance in the context of determining the number of common factors suggests it may often perform poorly (Lorenzo-Seva et al., 2011).

References

Allen, S. J., & Hubbard, R. (1986). Regression equations for latent roots of random data correlation matrices with unities on the diagonal. *Multivariate Behavioral Research, 21*, 393–398.

Briggs, N. E., & MacCallum, R. C. (2003). Recovery of weak common factors by maximum likelihood and ordinary least squares estimation. *Multivariate Behavioral Research, 38,* 25–56.

Browne, M. W. (2001). An overview of analytic rotation in exploratory factor analysis. *Multivariate Behavioral Research, 36,* 111–150.

Browne, M. W., & Cudeck, R. (1992). Alternative ways of assessing model fit. *Sociological Methods and Research, 21,* 230–258.

Browne, M. W., Cueck, R., Tateneni, K., & Mels, G. (2010). CEFA: Comprehensive Exploratory Factor Analysis. Version 3.04 [Computer software and manual]. Retrieved from http://faculty.psy.ohio-state.edu/browne/

Buja, A., & Eyuboglu, N. (1992). Remarks on parallel analysis. *Multivariate Behavioral Research, 27,* 509–540.

Carroll, J. B. (1953). An analytic solution for approximating simple structure in factor analysis. *Psychometrika, 18,* 23–28.

Cattell, R. B. (1966). The scree test for the number of factors. *Multivariate Behavioral Research, 1,* 245–276.

Cattell, R. B. (1978). *The scientific use of factor analysis in behavioral and life sciences.* New York, NY: Plenum.

Cattell, R. B., & Jaspers, J. (1967). A general plamode (No. 30-10-5-2) for factor analytic exercises and research. *Multivariate Behavioral Research Monographs,* No. 67-3.

Cattell, R. B., & Vogelmann, S. (1977). A comprehensive trial of the scree and KG criteria for determining the number of factors. *Multivariate Behavioral Research, 12,* 289–325.

Comrey A. L. (1978). Common methodological problems in factor analytic studies. *Journal of Consulting and Clinical Psychology, 46,* 648–659.

Cota, A. A., Longman, R. S., Holden, R. R., & Fekken, G. C. (1993). Comparing different methods of implementing parallel analysis: A practical index of accuracy. *Educational and Psychological Measurement, 53,* 865–876.

Cota, A. A., Longman, R. S., Holden, R. R., Fekken, G. C., & Xinaris, S. (1993). Interpolating 95th percentile eigenvalues from random data: An empirical example. *Educational and Psychological Measurement, 53,* 585-596.

Cudeck, R., & Henly, S. J. (1991). Model selection in covariance structures and the "problem" of sample size: A clarification. *Psychological Bulletin, 109,* 512–519.

Cudeck, R., & O'Dell, L. L. (1994). Applications of standard error estimates in unrestricted factor analysis: Significance tests for factor loadings and correlations. *Psychological Bulletin, 115,* 475–487.

Everitt, B. S. (1975). Multivariate analysis: The need for data and other problems. *British Journal of Psychiatry, 126,* 237–240.

Fabrigar, L. R., Visser, P. S., & Browne, M. W. (1997). Conceptual and methodological issues in testing the circumplex structure of data in personality and social psychology. *Personality and Social Psychology Review, 1,* 184–203.

Fabrigar, L. R., Wegener, D. T., MacCallum, R. C., & Strahan, E. J. (1999). Evaluating the use of exploratory factor analysis in psychological research. *Psychological Methods, 4,* 272–299.

Ferguson, G. A. (1954). The concept of parsimony in factor analysis. *Psychometrika, 19,* 347–362.

Finch, J. F., & West, S. G. (1997). The investigation of personality structure: Statistical models. *Journal of Research in Personality, 31*, 439-485.

Ford, J. K., MacCallum, R. C., & Tait, M. (1986). The applications of exploratory factor analysis in applied psychology: A critical review and analysis. *Personnel Psychology, 39*, 291–314.

Glorfeld, L. W. (1995). An improvement on Horn's parallel analysis methodology for selecting the correct number of factors to retain. *Educational and Psychological Measurement, 55*, 377–393.

Gorsuch, R. L. (1980). Factor score reliabilities and domain validities. *Educational and Psychological Measurement, 40*, 895–897.

Guttman, L. (1954). Some necessary conditions for common factor analysis. *Psychometrika, 19*, 149–161.

Guttman, L. (1956). "Best possible" systematic estimates of communality. *Psychometrika, 21*, 273–285.

Hakstian, A. R., Rogers, W. T., & Cattell, R. B. (1982). The behavior of number of factor rules with simulated data. *Multivariate Behavioral Research, 17*, 193–219.

Harman, H. H. (1976). *Modern factor analysis* (3rd edition). Chicago, IL; University of Chicago Press.

Harris, C. W., & Kaiser, H F. (1964). Oblique factor analytic transformations by orthogonal transformations. *Psychometrika, 29*, 347–362.

Harris, M. L., & Harris, C. W. (1971). A factor analytic interpretation strategy. *Educational and Psychological Measurement, 31*, 589–606.

Hendrickson, A. E., & White, P. O. (1964). PROMAX: A quick method for rotation to oblique simple structure. *British Journal of Statistical Psychology, 17*, 65–70.

Horn, J. L. (1965). A rationale and test for the number of factors in factor analysis. *Psychometrika, 30*, 179–185.

Horn, J. L. (1969). On the internal consistency reliability of factors. *Multivariate Behavioral Research, 4*, 115–125.

Hu, L. & Bentler, P. M. (1998). Fit indices in covariance structure modeling: Sensitivity to underparamterized model misspecifaction. *Psychological Methods, 3*, 424–453.

Humphreys, L. G., & Ilgen, D. R. (1969). Note on a criterion for the number of common factors. *Educational and Psychological Measurement, 29*, 571–578.

Humprheys, L. G., & Montanelli, R. G. (1975). An investigation of the parallel analysis criterion for determining the number of common factors. *Multivariate Behavioral Research, 10*, 193–206.

Jenrich, R. I., & Sampson, P. F. (1968). Rotation for simple loadings. *Psychometrika, 31*, 313–323.

Kaiser, H. F. (1958). The varimax criterion for analytic rotation in factor analysis. *Psychometrika, 23*, 187–200.

Kaiser, H. F. (1970). The application of electronic computers to factor analysis. *Educational and Psychological Measurement, 20*, 141–151.

Lautenschlager, G. J. (1989). A comparison of alternatives to conducting Monte Carlo analyses for determining parallel analysis criteria. *Multivariate Behavioral Research, 24*, 365–395.

Lautenschlager, G. J., Lance, C. E., & Flaherty, V. L. (1989). Parallel analysis criteria: Revised equations for estimating latent roots of random data correlation matrices. *Educational and Psychological Measurement, 49*, 339–345.

Lawley, D. N. (1940). The estimation of factor loadings by the method of maximum likelihood. *Proceedings of the Royal Society of Edinborough, 60A*, 64–82.

Linn, R. L. (1968). A Monte Carlo approach to the number of factors problem. *Psychometrika, 33*, 37–72.

Longman, R. S., Cota, A. A., Holden, R. R., & Fekken, G. C. (1989). A regression equation for the parallel anaysis criterion in principal components analysis: Mean and 95th percentile eigenvalues. *Multivariate Behavioral Research, 24*, 59–69.

Lorenzo-Seva, U., & Ferrando, P. J. (2006). FACTOR: A computer program to fit the exploratory factor analysis model. *Behavior Research Methods, 38*, 88–91.

Lorenzo-Seva, U., Timmerman, M. E., & Kiers, H. A. L. (2011). The hull method for selecting the number of common factors. *Multivariate Behavioral Research, 46*, 340–364.

MacCallum, R. C. (1990). The need for alternative measures of fit in covariance structure modeling. *Multivariate Behavioral Research, 25*, 157–162.

MacCallum, R. C., Browne, M. W., & Cai, L. (2007). Factor analysis models as approximations. In R. Cudeck & R. C. MacCallum (Eds.), *Factor analysis at 100: Historical developments and future directions* (pp. 153–175). Mahwah, NJ: Lawrence Erlbaum Associates.

MacCallum, R. C., Browne, M. W., & Sugawara, H. M. (1996). Power analysis and determination of sample size in covariance structure modeling. *Psychological Methods, 1*, 130–149.

Montanelli, R. G., & Humphreys, L. G. (1976). Latent roots of random data correlation matrices with squared multiple correlations on the diagonals: A Monte Carlo study. *Psychometrika, 41*, 341–348.

Neuhaus,, J. O., & Wrigley, C. (1954). The quartimax method: An analytical approach to orthogonal simple structure. *British Journal of Statistical Psychology, 7*, 187–191.

O'Connor, B. P. (2000). SPSS and SAS programs for determining the number of components using parallel analysis and Velicer's MAP test. *Behavioral Research Methods, Instruments, and Computers, 32*, 396–402.

Sass, D. A., & Schmitt, T. A. (2010). A comparative investigation of rotation criteria within exploratory factor analysis. *Multivariate Behavioral Research, 45*, 73–103.

Saunders, D. R. (1953). *An analytic method for rotation to orthogonal simple structure*. Research Bulletin, 53-10, Princeton, NJ: Educational Testing Service.

Steiger, J. H. & Lind, J. (1980, May). *Statistically based tests for the number of common factors*. Paper presented at the annual meeting of the Psychometric Society, Iowa City, IA.

Thompson, J. (2004). *A Monte Carlo comparison of tests for the number of factors under alternative factor models*. Unpublished doctoral dissertation, University of California, Davis.

Thurstone, L. L. (1935). *The vectors of the mind; multiple-factor analysis for the isolation of primary traits*. Chicago, IL: University of Chicago Press.

Thurstone, L. L. (1947). *Multiple-factor analysis; A development and expansion of The vectors of the mind.* Chicago, IL: University of Chicago Press.

Timmerman, M. E., & Lorenzo-Seva, U. (2011). Dimensionality assessment of ordered polytomous items with parallel analysis. *Psychological Methods, 16,* 209–220.

Tucker, L. R., Koopman, R. F., & Linn, R. L. (1969). Evaluation of factor analytic research procedures by means of simulated correlation matrices. *Psychometrika, 34,* 421–459.

Velicer, W. F. (1976). Determining the number of components from the matrix of partial correlations. *Psychometrika, 41,* 321–327.

Widaman, K. F. (1993). Common factor analysis versus principal component analysis: Differential bias in representing model parameters? *Multivariate Behavioral Research, 28,* 263–311.

Wood, J. M., Tataryn, D. J., & Gorsuch, R. L. (1996). Effects of under- or overextraction on principal axis factor analysis with varimax rotation. *Psychological Methods, 1,* 354–365.

Zwick, W. R., & Velicer, W. F. (1982). Factors influencing four rules for determining the number of components to retain. *Multivariate Behavioral Research, 17,* 253–269.

Zwick, W. R., & Velicer, W. F. (1986). Comparison of five rules for determining the number of components to retain. *Psychological Bulletin, 99,* 432–442.

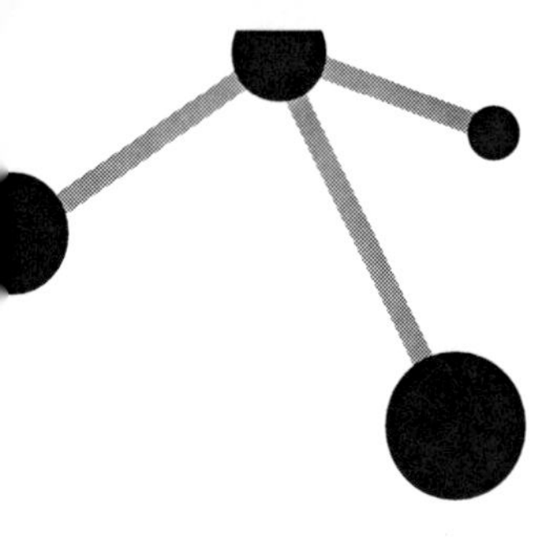

4

FACTOR ANALYSIS ASSUMPTIONS

IN CHAPTERS 1–3, we briefly touched upon various assumptions underlying the common factor model and the procedures typically used in its implementation. The goal for chapter 4 is to revisit these assumptions in more detail than was possible in the previous chapters. We begin our discussion with key assumptions underlying the common factor model itself, most notably with respect to its assumptions about how common factors influence measured variables. We then move to a discussion of assumptions underlying various procedures typically used to fit the common factor model to data. In discussing both sets of assumptions, our focus is threefold. First, we explain the nature of each assumption and discuss when it is or is not likely to be plausible. We then discuss methods for evaluating the plausibility of the assumption. We conclude with comments on various courses of action when a given assumption is not met.

Assumptions Underlying the Common Factor Model

Effects Indicator Models Versus Causal Indicator Models

Readers may recall from our discussion in chapter 1 that the common factor model assumes that scores on each measured variable in a battery are a result of one or more common factors as well as a unique factor. Thus, as illustrated in figure 4.1, the common factor model postulates that common factors exert linear causal effects on measured variables. This property makes the common factor model a member of a larger class of models often referred to as effects indicator models (e.g., see Bollen, 1989; Bollen & Lennox, 1991; Edwards & Bagozzi, 2000; MacCallum & Browne, 1993). Such models assume that latent variables cause measured variables and are called effects indicator models because they postulate that the measured variables (i.e., indicators) are effects of the common factors (i.e., latent variables). Such models are also sometimes referred to as reflective-measure models because the measured variables are presumed to reflect the common factors underlying them. This basic assumption of common factors (i.e., unobservable constructs) causing measures is not only central to the common factor model, but it also forms the foundation of classical test theory (Lord & Novick, 1968). As noted in chapter 1, models assuming that factors cause measured variables imply that measured variables strongly influenced by the same underlying

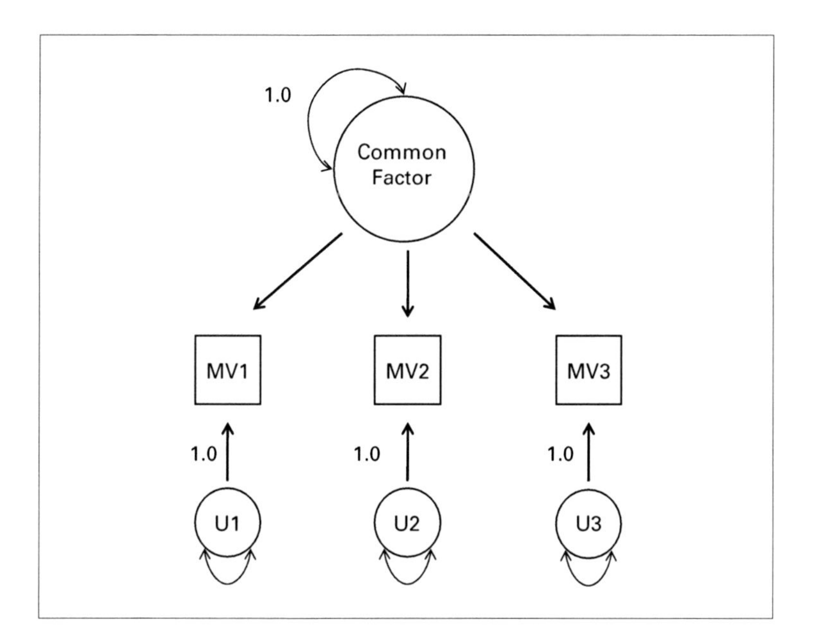

Figure 4.1. Graphical Representation of an Effects Indicator Model for an Example with One Common Factor and Three Measured Variables

factors should be substantially correlated with one another. Also, directly following from this logic, a battery of measured variables that all are all strongly influenced by the same underlying factors would be expected to produce high levels of internal consistency (e.g., high Cronbach alpha scores).

The assumption of common factors causing measured variables is quite plausible for many constructs. For example, the attitude construct lends itself readily to such an assumption. If a person has an attitude toward a given object (i.e., a positive or negative general evaluation of an object), it is conceptually plausible to postulate that this general evaluation would exert a causal influence on the extent to which people endorsed verbal statements implying positive or negative responses regarding the object. Likewise, the existence of an underlying personality trait such as extroversion would likely exert a causal influence on the degree to which people endorse verbal statements reflecting an interest in social interactions as self-descriptive. Similarly, it is sensible to assume that cognitive ability constructs such as math proficiency should exert a causal influence on the extent to which a person can successfully complete a series of math problems. Thus, across a wide range of constructs commonly studied in psychology and other social sciences, the underlying assumption that common factors cause measured variables is entirely reasonable.

The plausibility of this assumption in many contexts notwithstanding, a number of methodologists have noted that it may not be sensible to assume such a relationship between constructs and their measures in all cases (see Bollen, 1989; Bollen & Lennox, 1991; Edwards & Bagozzi, 2000; MacCallum & Browne, 1993). Rather, for some constructs, it may be more conceptually sensible to postulate the opposite causal assumption such that measured variables come together to constitute or cause the construct (see figure 4.2). Models based on this assumption are often referred to as *causal indicator models* (because indicators cause the construct) or as *formative-indicator models* (because the construct is formed from the indicators). As is illustrated in figure 4.2, causal indicator models postulate that the construct is a linear combination of the measured variables as well as some residual source of variance.

Methodologists have highlighted a number of constructs that might be more productively conceptualized in this manner. For example, socioeconomic status (SES) is often measured using indicators such as level of education, occupational prestige, and income.

An effects indicator model would assume that increases in some underlying quality in people, called SES, causes people to obtain high levels of education, work in prestigious occupations, and earn substantial income. It is not clear that postulating the existence of such an unobservable construct is compelling in this case. Instead, it may be more reasonable to argue that a person has high SES to the extent that they are able to obtain a high level of education, work in a prestigious occupation, and earn a high income. That is, the extent to which a person scores high on each of these criteria determines the degree to which they are high in SES. Other commonly studied constructs that might also be usefully conceptualized in terms of a causal indicator model could include constructs such as prevalence of life stressors and employee workload.

It is important to note that causal indicator models such as the model depicted in figure 4.2 have somewhat different implications regarding properties of measured variables. Perhaps most notably, causal indicator models do not imply that the measured variables comprising a battery should be highly correlated with one another. These models typically permit indicators to be correlated (as is depicted in figure 4.2), but the indicators themselves would only be expected to be correlated with one another to the extent that they share common antecedents. Because no antecedents are explicitly specified in the model, it leaves open the possibility that the

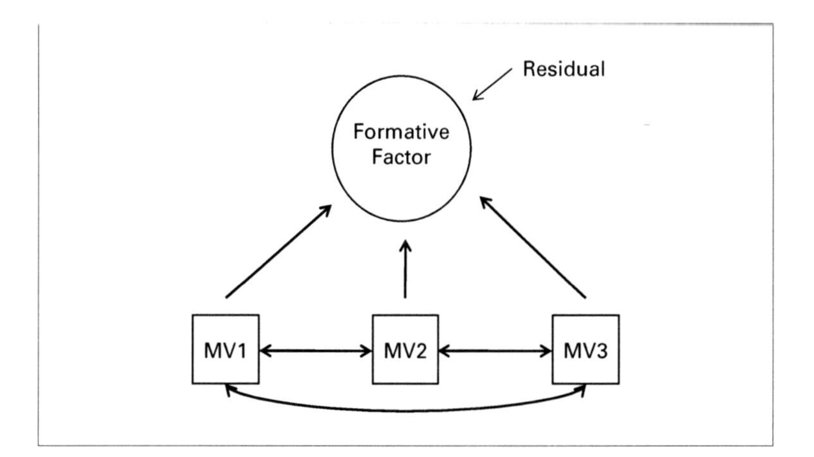

Figure 4.2. Graphical Representation of a Causal Indicator Model for an Example with One Factor and Three Measured Variables

indicators might have few if any shared antecedents. One logical extension from this fact is that one would also not necessarily expect scales constructed from such measures to have high levels of internal consistency. Of course, causal indicator models do not preclude a scale of such items manifesting other forms of reliability such as test-retest reliability.

Ultimately, the distinction between effects indicator models and causal indicator models speaks to the fundamental appropriateness of EFA in a given research context. EFA is only appropriate in contexts in which it is reasonable to assume an effects indicator model. The whole purpose of EFA is to arrive at a more parsimonious representation of the structure of correlations among measured variables by arriving at a comparatively small number of latent variables that can account for the pattern of correlations among measured variables. Causal indicator models make no clear assumptions about how measured variables comprising a given construct should be related to one another or to measured variables comprising other constructs, so such models cannot really be used to address the sorts of questions for which EFA is typically employed.

Thus, before undertaking EFA, it is essential to first consider the plausibility of an effects indicator model within the specific research context. Such a consideration must primarily be addressed on the basis of conceptual arguments (see Edwards & Bagozzi, 2000). That is, the researcher contemplating the use of EFA should carefully consider the nature of the constructs that would be expected to emerge in the domain of interest. If the sorts of constructs that are expected can be reasonably postulated to cause scores on measured variables, then EFA might well be a viable method of analysis. On the other hand, if a causal indicator model is more plausible, EFA should not be used. At the empirical level, there are really no clear exploratory methods for researchers to gauge the plausibility of their assumption of an effects indicator model in the context of EFA (although tests have been proposed for confirmatory latent variable models, see Bollen & Ting, 2000). Strong correlations among all measured variables or subsets of measured variables within a battery are certainly consistent with an effects indicator model. However, as noted, causal indicator models do not preclude high correlations among measured variables.

Conversely, failure to observe substantial correlations among measured variables comprising a battery might well indicate that a causal indicator model is more appropriate than an effects indicator model. Thus, if no parsimonious set of common factors that adequately accounts for the data can be arrived at, this outcome could be a result of the data reflecting a causal indicator model structure. However, it is important to recognize that this is not the only explanation for such an outcome. For example, an effects indicator model could still be appropriate but the measured variable might contain such high levels of random error that high correlations among the measured variables are not observed. Likewise, it might be difficult to arrive at a parsimonious set of common factors that can adequately account for the data because the domain lacks strong common factors or because the measured variables have been poorly sampled from the domain of interest. Thus, although patterns in the structure of correlations can be suggestive of the appropriateness of either model, they seldom provide a compelling basis on their own to argue for either type of model.

A final issue that should be highlighted when discussing effects indicator versus causal indicator models is that the very concept of causal indicator models is itself not without some controversy. There has been substantial debate among researchers about the conceptual and practical utility of causal indicator models (e.g., see Bagozzi, 2007; Bollen, 2007; Howell, Breivik, & Wilcox, 2007a, 2007b). Some have maintained causal indicator models are appropriate in certain contexts. Others have argued these models have fundamental limitations and that, frequently, the use of such models occurs in contexts in which constructs have been poorly measured. These critics suggest that if better measurement practices are employed, effects indicator models might prove to be superior representations of the data, even for constructs that have traditionally been studied in causal indicator models. Regardless of the stand one takes on the merits of causal indicator models, the bottom line for conducting EFA is that researchers should always consider whether the assumption that common factors exert causal influences on measured variables is appropriate for the battery of measured variables being examined. If it is not, this suggests either that a fundamentally different assumption is appropriate regarding the directionality of the relations of factors to measured variables or that the battery of measured variables is a poor representation of the domain of interest. In either case, EFA is not appropriate for the data at hand.

Linear versus Nonlinear Effects of Common Factors

A second key assumption of the common factor model is that common factors exert linear effects on measured variables. Stated another way, the common factor model assumes that each measured variable is a weighted linear combination of the common factors underlying the battery and a unique factor. Statistical models assuming linear effects are quite common in psychology and other social sciences. In many cases, assuming linear effects may be an accurate representation of the data or at least a sufficiently close approximation of the data so as not to introduce substantial distortions.

That being said, there are clear contexts in which assuming linear relations between common factors and measured variables might not be tenable. Perhaps the most obvious example of such a context is when the measured variables have only nominal or ordinal scale properties. Recall that a factor loading reflects the slope of increase (or decrease if the loading is negative) in units of a measured variable for each unit of increase in the common factor. Obviously, such an index of association between a common factor and a measured variable can only be meaningful if the units of the measured variable can be ordered and the distance between these units are equal. Thus, the common factor model is usually regarded as appropriate only in contexts in which measured variables have an interval or ratio level scale of measurement (e.g., see Gorsuch, 1983) or at least a "quasi-interval" level of measurement (Floyd & Widaman, 1995). Hence, nominal or ordinal measured variables with three or more categories will not satisfy the assumption of linear effects of common factors on measured variables.

Dichotomous measured variables constitute a somewhat different deviation from the interval or ratio scales of measurement normally assumed to be necessary to satisfy the assumption of linear effects of common factors on measured variables (see Floyd & Widaman, 1995; Gorsuch, 1983). Although dichotomous measures can be factor analyzed using standard EFA techniques, results of these analyses can sometimes be misleading. One central problem with the factor analysis of dichotomous items is that such items can produce "difficulty factors" that reflect variations in the endorsement rate of measured variables rather than the underlying constructs being assessed by the measured variables.

Because of this problem, most methodologists recommend that specialized factor analytic procedures be used for dichotomous measures.

Although an interval or ratio scale of measurement is usually necessary to meet the linearity assumption, such measurement properties do not guarantee that the assumption will be true. Common factors may sometimes have nonlinear effects on interval or ratio level measured variables. One situation in which nonlinearity might arise occurs when a set of measured variables are more effective in capturing variations in the underlying common factor at some levels of the factor than others. For example, imagine if a set of items had a comparatively high threshold to detect variations in the common factor of interest. In such a situation, increases at the low levels of the common factor would produce little or no change in the measured variables, but increases at the moderate and high levels of the common factor would produce changes in the measured variables. A relation of this sort would reflect both linear and curvilinear components. Likewise, situations in which the measured variables are incapable of capturing changes at high levels of the underlying common factor or situations in which the measured variables are incapable of capturing changes at both extremes would also produce associations that contain both a linear and curvilinear component. If the nonlinear component in these associations is comparatively weak relative to the linear component (e.g., the measured variables are only slightly less effective at capturing variations in the common factor at low levels than at moderate or high levels), the common factor model might still provide a reasonably good approximation of the data. However, if a strong nonlinear component is present, the common factor model might be a poor representation of the data.

A second general reason that nonlinearity might arise is if the common factors do not combine in an additive fashion to influence measured variables but, instead, interact with one another to influence measured variables. It is the norm for factor analytic models (and SEM more generally) to assume additive effects of common factors on measured variables. However, there are contexts in which methodologists have argued that it might be more appropriate to assume interactive effects among factors. For instance, some methodologists have suggested that data involving multitrait-multimethod measures (i.e., data in which each construct is assessed using each of a set of measurement procedures; Campbell & Fiske, 1959) might be more appropriately represented with models in which trait factors and method factors interact with one another rather than combining in an additive fashion to influence measured variables (e.g., see Browne, 1984; Campbell & O'Connell, 1967).

Determining when the assumption of linear effects of common factors on measured variables is or is not satisfied can be a comparatively straightforward or relatively difficult task depending on a number of methodological and conceptual circumstances. Contexts in which measured variables reflect only nominal or ordinal properties can often be comparatively easy to identify simply by examining the nature of the response options provided in the measured variables. For instance, response options that cannot be unambiguously ordered along some underlying conceptual continuum clearly are nominal in nature. Measures involving rank ordering also clearly make no attempt to gauge magnitude of differences between units and thus can reflect only ordinal level measurement. Traditional rating scales, on the other hand, at least in principle, can reflect interval level measurement. Generally, such scales have been regarded as sufficiently close to approximating interval level measurement that factor analysis can be considered as a means of analysis. Of course, in the case of rating scales or even measures that we can be confident have ratio level scales, there is no way to be certain that such items are equally sensitive at detecting variations in the underlying common factors at all ranges of these factors or that the underlying common factors combine within one another in a purely additive manner to influence measured variables.

That being said, there are some potential warning signs that often emerge if substantial nonlinear effects exist. One potential warning sign is poor model fit. The existence of strong nonlinear effects is one potential source of lack of model fit. Of course, poor fit can also result from other sources such as the failure to include an adequate number of common factors in the model. However, in contexts in which model fit is poor and there is no evidence of underfactoring, the existence of nonlinear effects should be seriously considered by the researcher. Another potential warning sign of nonlinear effects can be implausible or difficult to interpret parameter estimates (i.e., factor loadings and unique variances). Once again, aberrant parameter estimates can arise for a number of reasons including the model having an insufficient number of common factors. However, when they do arise, nonlinear effects should certainly be considered as one possible source, particularly if there is little indication that alternative explanations are responsible.

Assuming that a researcher does conclude that nonlinear effects are present, there are several options available to deal with this challenge. In some cases, it may be possible to transform measured variables so that linear associations between measured variables and common factors can be expected to emerge (see Gorsuch, 1983). For situations involving dichotomous measured variables, some researchers have explored strategies in which several dichotomous items are combined into "item parcels" so as to form a smaller set of measured variables that more closely approximate interval level measurement (e.g., Cattell, 1956; Cattell, & Bursdal, 1975; Gorsuch & Yagel, 1981 as cited in Gorsuch, 1983). One obvious challenge of this second approach is that it may not always be easy, especially in an exploratory context, to determine which items should be combined with one another to construct the various item parcels (see Little, Cunningham, Shahar, & Widaman, 2002). Also, such approaches require a comparatively large number of measured variables in the original battery. Finally, even when a clear basis exists for constructing item parcels, analyses based on different combinations of items to form parcels (even when alternative item combinations all assess the same construct) can produce notable differences in results if sample sizes are small and/or communalities are low (Sterba & MacCallum, 2010). Another potential response to dealing with nonlinear effects is to use a nonlinear factor analytic model. Substantial progress has been made in developing nonlinear factor analysis models (see Wall & Amemiya, 2007). These nonlinear factor analytic models cannot generally be implemented using traditional EFA computer programs, but instead require SEM programs or specialized software.

Assumptions Related to Model Fitting Procedures

In the prior sections, we reviewed assumptions underlying the common factor model. These are assumptions inherent in the model, and, as such, they are made regardless of the specific procedures a researcher uses to implement the model. However, there are other assumptions regarding the measured variables that are closely associated with all or some of the specific model fitting procedures used to implement the model rather than the model itself. Users of EFA must also be attentive to these assumptions.

Multivariate Normality

In chapter 3, we discussed the three most commonly used procedures for fitting the common factor model to the data. We noted that advantages of maximum likelihood (ML) model fitting include the ability to compute indices of model fit and to compute standard errors for parameter estimates (and by extension to conduct statistical tests of parameter estimates). However, we also highlighted that these advantages came with the additional assumption that the measured variables have a multivariate normal distribution.

One issue that has been of central interest to methodologists is how robust ML model fitting is to violations of this assumption when fitting latent variable models. Early in this literature, there was considerable concern regarding violations of normality assumptions. Fortunately, subsequent research has suggested that ML model fitting is more robust to violations of normality than was originally feared (e.g., Curran, West, & Finch, 1996). Such assurances, however, should not suggest that the issue can be ignored (see West, Finch, & Curran, 1995). Thus, when using ML EFA, two obvious questions arise. First, how does a researcher know when violations of normality are sufficiently severe to introduce substantial distortions in results? Second, when sufficiently severe violations have occurred, what remedies can be undertaken to deal with the problem?

A number of answers to the first question have been advanced. For example, many statistics textbooks provide general rules of thumb that can be applied to any statistical procedure assuming normality. Probably the most common of these general rules of thumb is to examine the ratios of skewness and kurtosis to their respective standard errors. Typically ratios of 2:1 or 3:1 are postulated to reflect severe nonnormality. General guidelines of this sort seem problematic for at least two reasons. First, they presume that all statistical procedures assuming normality are equally robust to violations of normality. Such an assumption is unlikely. Hence, any single recommendation is probably too conservative for some procedures and too liberal for others. A second objection is that these rules are very sensitive to sample size. Because standard errors will become smaller as sample sizes become larger, one implication of rules of this sort is that larger violations of normality are permitted in small sample sizes than in large sample sizes. The rationale for such a property is not at all clear. Given that most statistical estimation procedures tend to perform better in large samples than in small samples, if anything, one would argue that violations of normality should be more problematic in small sample sizes than large sample sizes.

We would argue that a more sensible approach to deriving guidelines for nonnormality is to examine the performance of a given statistical procedure under varying levels of nonnormality and other conditions (e.g., properties of the data and model) to determine the level of violations of normality at which substantial distortions in results emerge. Fortunately, studies of this sort have been conducted for ML model fitting in the context of fitting latent variable models (e.g., Curran et al., 1996). Substantial distortions in ML parameter estimates did not emerge until measured variables had an absolute value of skew of two or greater and an absolute value of kurtosis of seven or greater. Thus, data sets with skew and kurtosis values substantially smaller than these guidelines are unlikely to present problems for ML EFA, whereas data sets at or above these values might be problematic.

When severe nonnormality exists, researchers have a number of options available to them (see West et al., 1995). One option is to re-express the measured variables in a manner that produces measures with more normal distributions. Often, the aggregate of several measures will produce a more normal distribution than the distributions of the individual items making up that score. However, as discussed in the context of item parcels with dichotomous measures, it can sometimes be difficult to know which items should be combined to form a given parcel and different parceling decisions even for measures of the same construct can lead to differences in results when sample sizes are small and/or communalities are low. Also, such an approach may not be feasible when there are comparatively few measured variables in the original battery.

A second method for re-expressing measured variables is to conduct a nonlinear transformation on the measured variables, such as a power function transformation. This approach, although useful, is not without its challenges. One problem is determining the specific transformation that is most appropriate for a given data set. Numerous transformations exist and no single transformation will be best in all contexts. Moreover, in some cases, there may be no transformation that produces sufficient improvement to satisfy assumptions of normality. Thus, it is important for researchers to check skewness and kurtosis values following transformation to ensure that the transformation has been successful. Another drawback to transformations is that they can often present interpretational problems because variables are no longer on their original scales. Such problems can be especially difficult when comparing results across studies.

Another general response to dealing with nonnormal data is not to transform measured variables but, instead, to simply use another model fitting procedure that does not assume multivariate normality. In the context of EFA, NIPA factor analysis and IPA factor analysis provide two such alternatives. Both methods are reasonable approaches to fitting the common factor model and have some advantages over ML EFA. That being said, neither approach currently permits researchers to compute formal indices of model fit or standard errors for model parameters. Some recent work has been conducted to develop related OLS procedures that will permit the computation of such information. However, these procedures also assume multivariate normality, and their robustness to violations of this assumption have yet to be explored (e.g., see Browne, Cudeck, Tateneni, & Mels, 2010). When there are questions about multivariate normality assumptions, however, it can often be useful to compare results obtained using ML EFA with one of the methods that does not assume multivariate normality. This can give the researcher a better idea of whether the distributional shape of the measured variables is distorting the ML EFA results.

Perfect Linear Dependencies among Measured Variables

Another property of measured variables (often not discussed in reviews of factor analysis) is that none of the measured variables comprising a battery should be a perfect linear function of other measured variables in the battery. That is, a given measured variable in the battery should not be perfectly accounted for by a linear combination of other measured variables or a linear transformation of another measured variable in the battery. One situation in which such a relation would exist is if a researcher included a measured variable in the battery that was the sum or average of several other measured variables in the battery. This aggregate measured variable would be perfectly explained by the measured variables making up that measure.

For cases in which such a variable exists in the battery, EFA model fitting procedures will be unable to arrive at a solution for the model parameters and will usually report a warning that the matrix to be analyzed is "not positive definite." A detailed discussion of the mathematical reasons that such correlation matrices pose a problem is beyond the scope of this chapter. However, at the conceptual level, one way to think about this issue is that any factor analysis of a battery assumes that all measured variables in the battery contribute some unique information. When a measured variable is a perfect linear function of other variables in the model, this measured variable is, in effect, redundant with these other variables and, as such, provides no unique information. Thus, as a general practice, researchers should avoid inclusion of measured variables that are linear aggregates of other measured variables in the analysis or are linear transformations of another measured variable in the analysis.

Summary and Conclusions

We have now reviewed the key assumptions underlying the common factor model and the procedures used to fit this model to data. Ideally, researchers should carefully consider these assumptions prior to collecting any data for which an EFA is likely to be used. The reason for being attentive to such assumptions is that several of the assumptions derive in part from properties of the measured variables. Thus, a researcher's choices in the selection and/or development of the measured variables can have a major impact on the degree to which some assumptions are satisfied. Once the data have been collected, researchers should also carefully consider these assumptions and be attentive to both conceptual and empirical criteria by which the plausibility of assumptions can be gauged.

References

Bagozzi, R. P. (2007). On the meaning of formative measurement and how it differs from reflective measurement: Comment on Howell, Breivik, and Wilcox (2007). *Psychological Methods, 12*, 229–237.

Bollen, K. A. (1989). *Structural equations with latent variables.* New York: Wiley.

Bollen, K. A. (2007). Interpretational confounding is due to misspecifaction, not to type of indicator: Comment on Howell, Breivik, and Wilcox (2007). *Psychological Methods, 12*, 219–228.

Bollen, K., & Lennox, R. (1991). Conventional wisdom on measurement: A structural equation perspective. *Psychological Bulletin, 110*, 305–314.

Bollen, K. A., & Ting, K. (2000). A tetrad test for causal indicators. *Psychological Methods, 5*, 3–32.

Browne, M. W. (1984). The decomposition of multitrait-multimethod matrices. *British Journal of Mathematical and Statistical Psychology, 37*, 1–21.

Browne, M. W., Cueck, R., Tateneni, K., & Mels, G. (2010). CEFA: Comprehensive Exploratory Factor Analysis. Version 3.04 [Computer software and manual]. Retrieved from http://faculty.psy.ohio-state.edu/browne/

Campbell, D. T., & Fiske, D. W. (1959). Convergent and discriminant validation by the multitrait-multimethod matrix. *Psychological Bulletin, 56*, 81–105.

Campbell, D. T., & O'Connell, E. J. (1967). Method factors in multitrait-multimethod matrices: Multiplicative rather than additive? *Multivariate Behavioral Research, 2*, 409–426.

Cattell, R. B. (1956). Validations and intensification of sixteen personality factor questionnaire. *Journal of Clinical Psychology, 12*, 205–212.

Cattell, R. B., & Bursdal, C. A. (1975). The radial parcel double factoring design: A solution to the item-vs-parcel controversy. *Multivariate Behavioral Research, 10*, 165–179.

Curran, P. J., West, S. G., & Finch, J. F. (1996). The robustness of test statistics to nonnormality and specification error in confirmatory factor analysis. *Psychological Methods, 1*, 16–29.

Edwards, J. R., & Bagozzi, R. P. (2000). On the nature and direction of relationships between constructs and measures. *Psychological Methods, 5*, 155–174.

Floyd, F. J., & Widaman, K. F. (1995). Factor analysis in the development and refinement of clinical assessment instruments. *Psychological Assessment, 7*, 286–299.

Gorsuch, R. (1983). *Factor analysis* (2nd ed.). Hillsdale, NJ: Erlbaum.

Howell, R. D., Breivik, E., & Wilcox, J. B. (2007a). Reconsidering formative measurement. *Psychological Methods, 12*, 205–218.

Howell, R. D., Breivik, E., & Wilcox, J. B. (2007b). Is formative measurement really measurement? Reply to Bollen (2007) and Bagozzi (2007). *Psychological Methods, 12*, 238–245.

Little, T. D., Cunningham, W. A., Shahar, G., & Widaman, K. F. (2002). To parcel or not to parcel: Exploring the question, weighing the merits. *Structural Equation Modeling, 9*, 151–173.

Lord, F. M., & Novick, M. R. (1968). *Statistical theories of mental test scores.* Reading, MA: Addison-Wesley.

MacCallum, R. C., & Browne, M. W. (1993). The use of causal indicators in covariance structure models: Some practical issues. *Psychological Bulletin, 114*, 533–541.

Sterba, S. K., & MacCallum, R. C. (2010). Variability in parameter estimates and model fit across repeated allocations of items to parcels. *Multivariate Behavioral Research, 45*, 322–358.

Wall, M. M., & Amemiya, Y. (2007). A review of nonlinear factor analysis and nonlinear structural equation modeling. In R. Cudeck & R. C. MacCallum (Eds.), *Factor analysis at 100: Historical developments and future directions* (pp. 337–361). Mahwah, NJ: Erlbaum.

West, S. G., Finch, J. F., & Curran, P. J. (1995). Structural equation models with nonnormal variables: Problems and remedies. In R. H. Hoyle (Ed.), *Structural equation modeling: Concepts, issues and applications* (pp. 56–75). Thousand Oaks, CA: Sage.

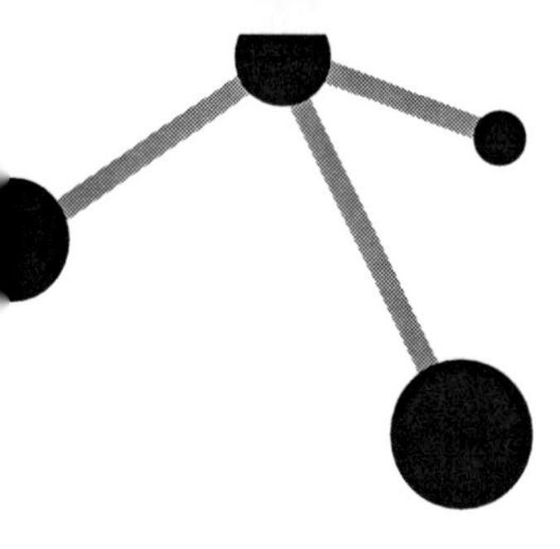

5

IMPLEMENTING AND INTERPRETING EXPLORATORY FACTOR ANALYSIS

IN CHAPTERS 1–4, we provided a conceptual overview of the common factor model, its underlying assumptions, and key procedural issues in its implementation. The goal of chapter 5 is to illustrate how many of the key procedures of an exploratory factor analysis (EFA) can be implemented in practice and how the information provided by an EFA can be interpreted. We begin with a brief review of key considerations that researchers should address prior to conducting an EFA. Next, we introduce the data set that will be used throughout the chapter to illustrate the implementation and interpretation of the EFA. We then move into our discussion of the EFA for the data set. We start with procedures for determining the appropriate number of factors. Next, we review program syntax for conducting an EFA with a specified number of common factors. We conclude our illustration with interpretation of key aspects of the output provided by widely used EFA programs. Throughout the chapter, we discuss the implementation and interpretation of EFA in the context of three programs that are frequently used to conduct EFA: SPSS, SAS, and CEFA.

Context for the Analysis: Pre-Analysis Decisions

For this chapter, we assume that one has already addressed a number of key steps (see chapter 2, Fabrigar et al., 1999, and Wegener & Fabrigar, 2004, for additional discussion of these pre-analysis steps). For example, we assume that one has determined the nature and scope of the research question. If one hopes to address the theoretical question of whether a certain set of constructs can account for the structure of correlations among a set of measured variables, we assume that the set of measured variables has already been chosen. Similarly, if one hopes to develop an effective set of measures to assess certain constructs, we assume that multiple potential items for each construct have been developed and that the researcher is ready to put those items to the test. Recall that selection of measures includes issues such as adequate sampling from the domain of interest and use of a sufficient number of measured variables to provide multiple assessments of each common factor (i.e., overdetermination of the common factors; see Fabrigar et al., 1999, for additional discussion). We also assume that we are dealing with scales of measurement appropriate to the EFA method used and that measures display appropriate distributional properties when using an EFA method that makes distributional assumptions (such as ML factoring).

Pre-analysis decisions also include the nature and size of the sample. Of course, as discussed in chapter 2, the adequacy of the sample size depends on a number of properties of the data and the model. For our example, which is based on a simulated data set, we specified a sample size of 200, which should be sufficient with moderate to high communalities and three or more items per factor. In this chapter, we conduct the example analyses using sufficient sample size rather than examining whether that sample is sufficiently large (see MacCallum et al., 1999, 2001, for additional discussion of determination of sample size in EFA). One would also want to choose a sample that adequately represents the population of interest as it relates to the domain under study. For our example addressing forms of social support, although the actual data set is artificially constructed, for purposes of discussion we assume that one is sampling from graduate students in social science disciplines at Big-Ten institutions in the midwestern United States. This sample might or might not represent the U.S. or world population as a whole, but it would seem to be a reasonable group for which the measures utilized would be meaningful and would vary substantially from person to person within the sample.

Example Research Question and Measures

Consider a hypothetical case in which one intends to explore different forms of social support that graduate students choose to give to friends and family (i.e., material social support and psychological social support). Let us imagine that we create a set of measures in which the person is asked to report how much of nine specific types of support she or he offers in an average week (using the previous three months as the time period of interest). The nine specific types of support are (1) the number of hugs the person gives (Hugs), (2) the number of compliments the person gives (Comps), (3) the number of times the person gives another person advice about his or her personal life (PerAd), (4) the number of times the person invites someone to social activities (SocAc), (5) the number of times the person provides some type of professional advice (ProAd), (6) how often the person participates in communal study sessions (i.e., studying in a group; ComSt), (7) the number of times the person provides some form of physical help, such as yard or house maintenance (PhyHlp), (8) how often the person explicitly encourages others (Encour), and (9) how often the person tutors other students on an academic subject (Tutor). Assume that responses from the 200 research participants create the simulated correlation matrix listed in table 5.1 (the simulated raw data can be downloaded from www.oup.com/us/exploratoryfactoranalysis).

For this example, we have already determined that we wish to examine the latent constructs underlying the pattern of correlations among our measured variables and, because we are exploring new measures rather than using an established scale, we opt for an EFA rather than a confirmatory factor analysis (CFA; though use of an existing scale does not necessarily mean that one must use a CFA). We also proceed through the rest of this chapter assuming that we have chosen an EFA rather than a PCA. It is worth noting that SPSS and SAS both have PCA as the default setting for their factor analysis routines. As was discussed in Chapter 3, PCA is not technically a factor analysis and is not widely recommended for situations in which researchers are interested in the

Table 5.1
Correlation Matrix for Example Data Set

	Hugs	Comps	PerAd	SocAc	ProAd	ComSt	PhyHlp	Encour	Tutor
Hugs	1.000								
Comps	0.666	1.000							
PerAd	0.150	0.247	1.000						
SocAc	0.617	0.576	0.222	1.000					
ProAd	0.541	0.510	0.081	0.409	1.000				
ComSt	0.653	0.642	0.164	0.560	0.667	1.000			
PhyHlp	0.473	0.425	0.091	0.338	0.734	0.596	1.000		
Encour	0.549	0.544	0.181	0.448	0.465	0.540	0.432	1.000	
Tutor	0.566	0.488	0.120	0.349	0.754	0.672	0.718	0.412	1.000

latent factors underlying relations among measured variables (as is often the case for disciplines in which there are substantive interests in identifying personal, social, institutional, or cultural factors that influence human behavior).

In the sections that follow, we describe how to conduct the EFA in three statistical packages that are frequently used to conduct factor analysis (i.e., SPSS, SAS, and CEFA). We also present output and note points of interest in how the procedure or output might differ across programs. In some cases, we will include calculations or macros that add to the information included in the standard packages or will point to alternative means of obtaining that desired information.

Conducting the Analysis: Implementation of EFA

In our example analyses, we will focus on the use of syntax. In many cases, the syntax presented can be created by pointing and clicking on drop-down menus and then opting to paste the syntax. However, use of syntax can be advantageous because there are certain options available in syntax that do not appear in the drop-down menus. One can often change a simple aspect of an analysis (such as the number of factors extracted) more quickly by replacing one number in syntax and then rerunning the analysis. Pointing and clicking to recreate the same analysis with that one change would often take much longer. In addition, we describe a number of SPSS or SAS macros that can quickly and easily provide useful (often needed) information. By necessity, those macros use syntax, so familiarity with the syntax for one's favorite program is a good idea. For example, none of the existing programs include parallel analysis as a drop-down option or as a documented procedure for determining the number of factors to extract. However, SPSS and SAS macros are available for this purpose, and we will illustrate their use.

Determining the Number of Factors

Usually the first thing a researcher must do in implementing an EFA is to determine the appropriate number of common factors. We discussed a variety of statistical procedures that have been used to aid in this decision. However, as we noted, some of these procedures are problematic (e.g., the eigenvalues-greater-than-one rule). Thus, in this section, we focus on the three procedures that we think provide the most useful information to assist in this decision: scree test, parallel analysis, and model fit. In reviewing these procedures, we emphasize the need to interpret the results in a holistic manner rather than interpreting each in isolation.

Scree Test. As discussed in chapter 3, one useful method for determining the appropriate number of common factors is to compute eigenvalues from the matrix to be analyzed and then plot these eigenvalues in descending order. The appropriate number of factors corresponds to the number of eigenvalues prior to the last major drop in the plot. In SPSS, the user can request a plot of eigenvalues. However, as noted in chapter 3, there are actually two sets of eigenvalues that have often been used as the basis for a scree test. Often, eigenvalues from the original (unreduced) correlation matrix (the matrix of correlations with 1's in the diagonal) are used. These eigenvalues are appropriate in the context of PCA, and they were the eigenvalues around which the eigenvalues-greater-than-one rule (for PCA) was originally developed. However, when conducting common factor analysis, the eigenvalues from the reduced matrix (the matrix of correlations with SMCs in the diagonal) are more appropriate. Therefore, one would typically want to obtain a scree plot based on eigenvalues from the reduced matrix.

Unfortunately, the SPSS factor analysis routine always reports eigenvalues from the unreduced matrix irrespective of whether PCA or common factor analysis has been conducted. Thus, when conducting an EFA in SPSS, it is necessary to run an SPSS macro (O'Connor, 2000) to compute the reduced matrix eigenvalues. The SPSS syntax for this macro in the context of our chapter 5 example is presented in table 5.2. The resulting eigenvalues produced by this macro appear in table 5.3. As the reader can see in figure 5.1, a scree plot of the eigenvalues includes one very large eigenvalue with a second eigenvalue that also stands out from the other eigenvalues (all of which are close to zero). This pattern stands in some contrast to the eigenvalues from the unreduced matrix (presented later in the chapter), where it looks more like a single factor (though the poorly performing eigenvalues-greater-than-one rule

Table 5.2
SPSS Syntax for Generating Eigenvalues from the Reduced Correlation Matrix (with kind permission from Springer Science + Business Media: Behavior Research Methods, Instruments, & Computers. "SPSS and SAS programs for determining the number of components using parallel analysis and Velicer's MAP test, 32, 2000, Brian P. O'Connor.

```
corr Hugs Comps PerAd SocAc ProAd ComSt PhyHlp
 Encour Tutor / matrix out ('f:\example1.sav') /
 missing = listwise.
matrix.
MGET /type= corr /file='f:\example1.sav' .
compute smc = 1 - (1 &/ diag(inv(cr)) ).
call setdiag(cr,smc).
compute evals = eval(cr).
print { t(1:nrow(cr)) , evals }
  /title="Raw Data Eigenvalues"
  /clabels="Root" "Eigen." /format "f12.6".
end matrix.
```

Note: First two lines of code use SPSS to generate a correlation matrix from a raw data set and save this matrix as an SPSS 'sav' file. The remaining lines of code read this correlation matrix into SPSS and compute eigenvalues from the reduced correlation matrix.

would happen to correctly identify the number of factors specified in the simulation because the second largest eigenvalue from the unreduced matrix is just larger than 1 whereas the third largest eigenvalue is just smaller than 1).

SAS differs a bit from SPSS in that SAS automatically produces the eigenvalues that go with the chosen model. Therefore, when a common factor model is chosen in SAS, the correct eigenvalues from the reduced matrix are printed. If a PCA is conducted in SAS, the eigenvalues from the unreduced matrix are printed. Unfortunately, CEFA does not print eigenvalues that can be used in a scree plot. It prints the eigenvalues from the sample (unreduced) matrix, but not those from the reduced matrix. Instead of using procedures focusing on eigenvalues, CEFA focuses on the use of model fit for determining the number of factors (discussed later in the chapter).

Table 5.3
SPSS Output of Eigenvalues from the Reduced Correlation Matrix Created using Syntax in Table 5.2

```
Matrix

[DataSet0]

Run MATRIX procedure:

MGET created matrix CR.
The matrix has 9 rows and 9 columns.
The matrix was read from the record(s) of row type
 CORR.

Raw Data Eigenvalues
         Root          Eigen
      1.000000      4.502823
      2.000000       .670159
      3.000000       .049239
      4.000000      -.002595
      5.000000      -.037097
      6.000000      -.046803
      7.000000      -.079295
      8.000000      -.116997
      9.000000      -.176831
------ END MATRIX -----
```

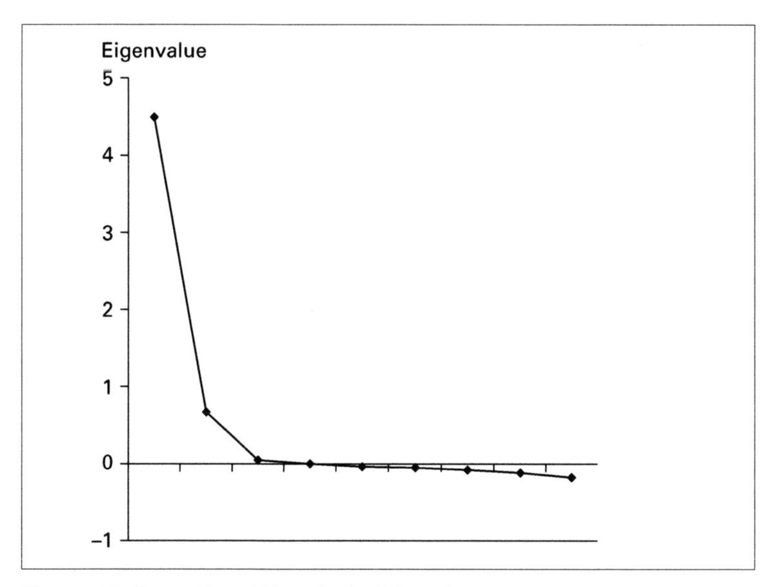

Figure 5.1. Scree Plot of Hypothetical Data Set

Parallel Analysis. A second factor number procedure that is also based on eigenvalues is parallel analysis (see chapter 3). Readers will recall that this procedure involves comparing eigenvalues from the reduced matrix (or unreduced matrix in the context of PCA) with eigenvalues that would be expected to emerge from a reduced matrix produced by random data. None of the three focal programs discussed in this chapter conducts a parallel analysis as a part of the regular program. However, SPSS and SAS macros have been developed to accomplish this goal (O'Connor, 2000). The SPSS syntax for the parallel analysis of our example is presented in table 5.4.

Of particular interest are the four lines of code under the heading "enter your specifications here" and the two lines of code under the heading "Specify desired kind of parallel analysis" near the beginning of the syntax file. These are the lines used to specify the settings for the parallel analysis. The "compute ncases" command specifies

Table 5.4

SPSS Syntax for a Parallel Analysis of a Reduced Correlation Matrix with Nine Variables Based on a Sample of 200 (with kind permission from Springer Science+Business Media: Behavior Research Methods, Instruments, & Computers. "SPSS and SAS programs for determining the number of components using parallel analysis and Velicer's MAP test, 32, 2000, pages 400–401, Brian P. O'Connor.

```
* Parallel Analysis program.

set mxloops=9000 printback=off width=80
 seed = 1953125.
matrix.

* enter your specifications here.
compute ncases = 200.
compute nvars = 9.
compute ndatsets = 100.
compute percent = 95.

* Specify the desired kind of parallel analysis,
 where:
  1 = principal components analysis
```

(*contd.*)

Table 5.4 (***contd.***)

```
  2 = principal axis/common factor analysis.
compute kind = 2 .

* principal components analysis.
do if (kind = 1).
compute evals = make(nvars,ndatsets,-9999).
compute nm1 = 1 / (ncases-1).
loop #nds = 1 to ndatsets.
compute x = sqrt(2 * (ln(uniform(ncases,nvars))
            * -1) ) &*
            cos(6.283185 * uniform(ncases,nvars)).
compute vcv = nm1 * (sscp(x) - ((t(csum(x))
              *csum(x))/ncases)).
compute d = inv(mdiag(sqrt(diag(vcv)))).
compute evals(:,#nds) = eval(d * vcv * d).
end loop.
end if.

* principal axis / common factor analysis with
 SMCs on the diagonal.
do if (kind = 2).
compute evals = make(nvars,ndatsets,-9999).
compute nm1 = 1 / (ncases-1).
loop #nds = 1 to ndatsets.
compute x = sqrt(2 * (ln(uniform(ncases,nvars))*
            -1) ) &*
            cos(6.283185 * uniform(ncases,nvars)).
compute vcv = nm1 * (sscp(x) - ((t(csum(x))
              *csum(x))/ncases)).
compute d = inv(mdiag(sqrt(diag(vcv)))).
compute r = d * vcv * d.
compute smc = 1 - (1 &/ diag(inv(r)) ).
call setdiag(r,smc).
compute evals(:,#nds) = eval(r).
end loop.
end if.

* identifying the eigenvalues corresponding to
 the desired percentile.
compute num = rnd((percent*ndatsets)/100).
```

(*contd.*)

Table 5.4 ***(contd.)***

```
compute results = { t(1:nvars), t(1:nvars),
 t(1:nvars) }.
loop #root = 1 to nvars.
compute ranks = rnkorder(evals(#root,:)).
loop #col = 1 to ndatsets.
do if (ranks(1,#col) = num).
compute results(#root,3) = evals(#root,#col).
break.
end if.
end loop.
end loop.
compute results(:,2) = rsum(evals) / ndatsets.

print /title="PARALLEL ANALYSIS:".
do if (kind = 1).
print /title="Principal Components".
else if (kind = 2).
print /title="Principal Axis / Common factor
analysis".
end if.
compute specifs = {ncases; nvars; ndatsets;
 percent}.
print specifs /title="Specifications for this
 Run:"
  /rlabels="Ncases" "Nvars" "Ndatsets" "Percent".
print results /title="Random Data Eigenvalues"
  /clabels="Root" "Means" "Prcntyle" /format
   "f12.6".

end matrix.
```

the number of observations for the real data set (200 in the present example) for which the parallel analysis will be conducted. The "compute nvars" refers to the number of measured variables being examined in the real data set (nine in the present example). The "compute ndatsets" command specifies the number of random data sets on which the parallel analysis will be based. In the present example, this value has been set to 100. Thus, the parallel analysis will generate 100 random data sets of nine variables based on 200 observations. The reduced matrix eigenvalues for each data set will be calculated and the program will report the average of each eigenvalue across the 100 random data sets. The "compute percent" command indicates the percentile value that will be reported in addition to the mean eigenvalues. In our example, this is specified at 95, which indicates that the ninety-fifth percentile for each eigenvalue across the 100 data sets will be reported in addition to the mean of each eigenvalue. Finally, the "compute kind" command indicates whether the parallel analysis will be based on the reduced matrix eigenvalues (designated by a value of 2) or the unreduced matrix eigenvalues (designated by a value of 1). In the present example, we have specified reduced matrix eigenvalues, consistent with the fact that we are conducting an analysis based on the common factor model.

Table 5.5 presents the results of the parallel analysis. The first column (with the heading Root) indicates the ordinal position of the eigenvalue. The second column (with the heading Means) indicates the mean value of that eigenvalue across the 100 random data sets. The final column (with the heading Prcntyle) reports the somewhat more stringent standard of the ninety-fifth percentile value of each eigenvalue for the 100 randomly generated data sets. Comparing the obtained eigenvalues from the reduced matrix (see table 5.3) with the random eigenvalues from the parallel analysis (see table 5.5), the two largest obtained eigenvalues from our actual data set exceed the random eigenvalues from the parallel analysis based on either the mean values or the ninety-fifth percentile values. Therefore, the parallel analysis suggests the same number of factors as the scree plot in this case.

Although both the mean and ninety-fifth percentile criteria of the parallel analysis suggest the same number of factors in the present example, it is, of course, possible for the two standards to produce different results. That is, in some cases, an observed eigenvalue from real data might exceed the mean value from random data but not the ninety-fifth percentile value. To date, there is no clear consensus in the methodological literature on which standard is best, though, to the extent that one is concerned that parallel analysis represents a relatively lenient standard for identifying "major" common factors, one might want to focus on the ninety-fifth percentile values. It is often useful to consider both criteria and recognize that diverging results between them simply highlight that there may be more than one plausible solution for the data.

Table 5.5

SPSS Output for a Parallel Analysis of a Reduced Correlation Matrix with Nine Variables Based on a Sample of 200 (with kind permission from Springer Science+Business Media: Behavior Research Methods, Instruments, & Computers. "SPSS and SAS programs for determining the number of components using parallel analysis and Velicer's MAP test, 32, 2000, Brian P. O'Connor.

```
Matrix
[DataSet0]

Run MATRIX procedure:

PARALLEL ANALYSIS:
Principal Axis / Common factor analysis

Specifications for this Run:
Ncases    200
Nvars       9
Ndatsets  100
Percent    95

Random Data Eigenvalues
      Root          Means       Prcntyle
   1.000000      .381096      .481432
   2.000000      .260848      .337430
   3.000000      .171785      .239584
   4.000000      .093392      .147084
   5.000000      .022890      .071884
   6.000000     -.041554     -.006859
   7.000000     -.103038     -.066351
   8.000000     -.172639     -.129117
   9.000000     -.254960     -.202242

------ END MATRIX -----
```

Moreover, as we noted in chapter 3, parallel analysis based on the reduced correlation matrix with SMCs in the diagonal has been found to overfactor in at least some contexts. In light of this fact, it might be useful for researchers to also examine the results of a parallel analysis based on minimum rank factor analysis, which uses a reduced correlation matrix with an alternative estimate of initial communalities (Timmerman & Lorenzo-Seva,

2011). Unfortunately, this procedure is not available in any of the three programs discussed in this chapter. It can only be computed in the program FACTOR (Lorenzo-Seva & Ferrando, 2006). A full illustration of the present example using FACTOR is available at www.oup.com/us/exploratoryfactoranalysis. Ultimately, a final choice of a solution should, as we have stressed in this chapter and in chapter 3, be based on the totality of information provided by different criteria for selecting the number of factors.

The SAS syntax for the parallel analysis is presented in table 5.6. Specifications are very similar to that of the SPSS file. Of course, the output of this analysis would be basically the same as that presented in table 5.5. In SAS, one can use the eigenvalues automatically printed as part of the factor analysis procedure to compare with the random eigenvalues from the parallel analysis (because SAS prints the eigenvalues from the reduced matrix when the common factor model is being fit to the data).

Table 5.6

SAS Syntax for a Parallel Analysis of a Reduced Correlation Matrix with Nine Variables and a Sample of 200 (with kind permission from Springer Science+Business Media: Behavior Research Methods, Instruments, & Computers. "SPSS and SAS programs for determining the number of components using parallel analysis and Velicer's MAP test, 32, 2000, page 401, Brian P. O'Connor.

```
/* Parallel Analysis program */

options nocenter nodate nonumber linesize=90;
 title;
proc iml;
reset noname; seed = 1953125;

/* enter your specifications here */
Ncases = 200;
Nvars = 9;
Ndatsets = 100;
percent = 95;

/* Specify the desired kind of parallel analysis,
 where:
   1 = principal components analysis
```

(contd.)

Table 5.6 ***(contd.)***

```
 2 = principal axis/common factor analysis */
kind = 2 ;

/* set diagonal to a column vector module */
start setdiag(matname,vector);
do i = 1 to nrow(matname);
do j = 1 to ncol(matname);
if (i = j) then; matname[i,j] = vector[i,1];
end;end;
finish;

/* row sums module */
start rsum(matname);
rsums =j(nrow(matname),1);
do rows = 1 to nrow(matname);
dumr = matname[rows,];
rsums[rows,1]=sum(dumr);
end;
return(rsums);
finish;

/* principal components analysis */
if kind = 1 then do;
/* computing random data correlation matrices &
 eigenvalues */
evals = j(nvars,ndatsets,-9999);
nm1 = 1 / (ncases-1);
do nds = 1 to ndatsets;
x = normal(j(ncases,nvars)) ;
vcv = nm1 * (t(x)*x - ((t(x[+,])*x[+,])/ncases));
d = inv(diag(sqrt(vecdiag(vcv))));
evals[,nds] = eigval(d * vcv * d);
end;
end;

/* principal axis/common factor analysis with
 SMCs on the diagonal */
if kind = 2 then do;
/* computing random data correlation matrices &
 eigenvalues */
```

(contd.)

Table 5.6 ***contd.***

```
evals = j(nvars,ndatsets,-9999);
nm1 = 1 / (ncases-1);
do nds = 1 to ndatsets;
x = normal(j(ncases,nvars)) ;
vcv = nm1 * (t(x)*x - ((t(x[+,])*x[+,])/ncases));
d = inv(diag(sqrt(vecdiag(vcv))));
r = d * vcv * d;
smc = 1 - (1 / vecdiag(inv(r)) );
run setdiag(r,smc);
evals[,nds] = eigval(r);
end;
end;

/* identifying the eigenvalues corresponding to
 the desired percentile */
num = round((percent*ndatsets)/100);
results = j(nvars,3,-9999);
s = 1:nvars;
results[,1] = t(s);
do root = 1 to nvars;
ranks = rank(evals[root,]);
do col = 1 to ndatsets;
if (ranks[1,col] = num) then do;
results[root,3] = evals[root,col];
col = ndatsets;
end;
end;
end;
results[,2] = evals[,+] / ndatsets;

print, "Parallel Analysis:";
if (kind = 1) then; print, "Principal
 Components";
if (kind = 2) then do;
print "Principal Axis / Common factor analysis";
print "Compare the random data eigenvalues below
 to the";
print "real-data eigenvalues that are obtained
 from a";
```

(contd.)

Table 5.6 **(*contd.*)**

```
print "Common factor analysis in which the # of
 factors";
print "extracted equals the # of variables/items,
 and the";
print "number of iterations is fixed at zero
 (maxiter=0), as in:";
print "proc factor data=trial priors=smc maxiter
 = 0; run; ";
print "Or use the 'rawpar.sas program' to obtain
 the ";
print "baseline real-data eigenvalues.";
end;
specifs = (ncases // nvars // ndatsets // percent);
rlabels = {"Ncases" "Nvars" "Ndatsets" "Percent"};
print/ "Specifications for this Run:",
specifs[rowname=rlabels];
clabels={"Root" "Means" "Prcntyle"};
print "Random Data Eigenvalues",
results[colname=clabels format=12.6];

quit;
```

Model Fit. A third very useful technique for determining the number of factors is to examine model fit. This approach is possible with ML factoring, but not with NIPA or IPA factoring. This is one of the reasons to use ML when possible (though one should always feel free to compare ML results with other common factor model fitting procedures such as NIPA and IPA). As noted in chapter 3, essentially any model fit index that is used in CFA and structural equation modeling can also be computed for an EFA model using ML. Unfortunately, the fit indices reported in the EFA routines of the major statistical programs are often quite limited. The only index of model fit printed by the factor analysis routine in SPSS is a Chi-square test of goodness of fit. SAS also reports the Chi-square goodness of fit test along with some other indices, including the Tucker-Lewis/Non-Normed Fit Index. CEFA also reports the chi-square goodness-of-fit test along with root mean square error of approximation (RMSEA) and expected cross-validation index (ECVI) (see Browne & Cudeck, 1992; Fabrigar et al., 1999).

Fortunately, because all major EFA programs report the chi-square along with the degrees of freedom for the model, it is possible to use this information to calculate a much larger variety of fit indices (see Hu & Bentler, 1998). As noted in chapter 3, one fit index that has received some attention in the context of EFA as a means of determining the appropriate number of factors is RMSEA (Browne & Cudeck, 1992; Steiger & Lind, 1980). This index is a useful indicator of the number of major factors, because it is comparatively easy to compute from the information available in most EFA routines, and it takes into account the parsimony of the model being fit to the data. RMSEA can be calculated by first computing the ML sample discrepancy function from the SPSS or SAS output (the ML sample discrepancy function = $\chi^2/[N-1]$). This sample discrepancy function can then be used in the RMSEA formulas provided in chapter 3 (see Equations 3.6 and 3.7) to obtain RMSEA. Alternatively, RMSEA can be directly calculated from the chi-square value using Equation 5.1:

$$RMSEA = \sqrt{\frac{\left(\chi^2/df\right)-1}{(n-1)}} \tag{5.1}$$

When the chi-square is less than the degrees of freedom, RMSEA is set to zero.

Alternatively, a model's RMSEA along with its confidence interval can be computed using a simple program called FITMOD (which can be downloaded from http://faculty.psy.ohio-state.edu/browne/software.php). This program requires that the user input the sample ML discrepancy function, degrees of freedom for the model, and the number of free parameters for the model. The number of free parameters for the model can be calculated by first determining the number of unique elements in the correlation matrix (i.e., the diagonal and either the lower or upper half of the off-diagonal) to which the model is being fit. The number of unique elements in a correlation matrix can be calculated using Equation 5.2:

$$\text{\# Unique Elements} = [(\text{number of variables}) \times (\text{number of variables} + 1)]/2 \quad (5.2)$$

The number of free parameters is then determined by Equation 5.3:

$$\text{\# parameters} = (\text{\# unique elements} - \text{model df}) \quad (5.3)$$

As discussed in chapter 3, RMSEA can be used to determine the appropriate number of factors by specifying a series of models, beginning with a one-factor model, and continuing with each subsequent model having one additional factor. The upper limit of the number of factors for the sequence of models is the point at which additional factors would no longer be conceptually useful or for which the model would approach a point at which it would have zero degrees of freedom. The appropriate number of factors is then judged by examining the RMSEA values for the sequence of models. The appropriate number of factors is that model in the sequence that ideally: (1) fits the data well, (2) fits substantially better than a model with one less factor, and (3) does not fit substantially worse than a model with one more factor.

Table 5.7 presents the RMSEA values and their 90 percent confidence intervals for models specifying 1–3 common factors fit to the example data. The RMSEA for the one-factor model indicates

Table 5.7
RMSEA Values for Factor Analysis Models using Maximum Likelihood Model Fitting

Model	RMSEA Point Estimate	90% Confidence Interval
1 Factor	.156	.133 to .180
2 Factor	.000	.000 to .056
3 Factor	.000	.000 to .063

a model with poor fit (RMSEA = 0.156), thereby suggesting that a single factor does a poor job accounting for the correlations among the measured variables. Addition of a second factor to the model provides a major improvement in the fit and indicates that this model does an excellent job accounting for the correlations among the measured variables (RMSEA = 0.000). Moreover, the addition of a third factor can provide no added benefit to model fit because the fit of the two-factor model is nearly perfect.

Also of interest in table 5.7 are the confidence intervals associated with each RMSEA value. The width of the confidence intervals gives one an idea of the precision of the point estimates. If the confidence intervals are wide, then one might not want to put much trust in the point estimates. Thus, with wide confidence intervals, even fairly sizable drops in RMSEA might not really reflect a substantially better model. In the present example, the shifts in RMSEA from the one-factor model to the two-factor model are quite substantial. Moreover, the magnitudes of these shifts are so large when compared to the confidence intervals that there is little doubt that adding the second factor constitutes a substantial improvement. Indeed, it is interesting to note that the confidence interval of the three-factor model is actually wider than that of the two-factor model indicating that some precision in estimating model fit is lost with the additional factor.

Thus, when all the information in table 5.7 is taken into account, it is clear that the model fit approach supports a two-factor model. As noted in chapter 3, one need not restrict the model fit approach to RMSEA. One could examine the pattern of fit indices for a variety of fit indices that can be calculated from the discrepancy function, chi-square, and degrees of freedom. On the www.oup.com/us/exploratoryfactoranalysis web site, we present values of alternative fit indices for models fit to our example data.

Summary. In the current case, each of the number-of-factor procedures correctly pointed to two factors. This was to be expected because these are some of the best performing of the various number-of-factor methods that have been proposed over the years, and we specifically created a simulated data set that had two major common factors. However, there will be occasions in which the different procedures point to different numbers of common factors. For example, parallel analysis and model fit using RMSEA each depend on rather different criteria for determining the number of factors. Because parallel analysis uses eigenvalues of random factors as a baseline for determining the number of factors (a relatively lenient standard), it might often point to a larger number of factors than RMSEA. Because RMSEA examines overall adequacy of model fit, taking into account model parsimony, this index would be less likely to suggest addition of minor factors that influence a small number of items, even if those minor factors are clearly nonrandom. Thus, when these procedures (and the scree test) agree, this is probably a relatively strong empirical indication that the identified number of factors is reasonable.

These empirical indicators of number of factors are only helpful to the extent that they result in models that produce meaningful, interpretable solutions. The interpretability of solutions can only be assessed on the basis of the rotated solution, so it is necessary for us to go farther into the analysis before this is characterized for the current solutions. However, perhaps especially when the various empirical indicators of number of factors conflict, one might want to look to issues of factor interpretability as a way to decide what the most appropriate number of factors might be. Also, as noted in chapter 3, when one has multiple data sets in which to examine the same factor structure, then stability of solutions can also be a valuable criterion for judging the appropriate number of factors.

Syntax for Conducting the EFA

The model fit information (as well as the other output to be discussed) is only produced when the EFA is run in one's software of choice. There are a number of choices for software to run an EFA. Perhaps the ones most currently popular with social scientists are SPSS and SAS. CEFA is a less known but useful program, especially for providing a wide variety of factoring methods, rotations, and fit indices. In the following sections, we will present syntax for each that conducts EFA using ML factor extraction and oblique rotations for the two factors identified in the various factor number procedures. Different programs have different rotation options, which might shift the results slightly (sometimes more substantially if the rotations differ in terms of whether they specify orthogonal rather than oblique rotations, see Fabrigar et al., 1999). We will present the results of oblique rotations here, because they are more realistic and do not constrain the results to have a certain level of correlation (or lack thereof) among factors, but results using alternative orthogonal rotations can be found at www.oup.com/us/exploratoryfactoranalysis.

SPSS. Table 5.8 presents the syntax used to conduct an ML EFA specifying two factors and using a direct quartimin oblique rotation (specified in SPSS as a direct oblimin rotation). Note that the first set of commands reads in the data set (in this case a text file containing the correlation matrix) and specifies the variable names, location of the file, sample size, and nature of the entered data. This program reads in the correlation matrix directly. For syntax that runs on raw data to create the correlation matrix and complete the EFA, see the web site that accompanies the book (www.oup.com/us/exploratoryfactoranalysis). After the FACTOR

Table 5.8

SPSS Syntax for a Maximum Likelihood EFA with Two Common Factors and a Direct Quartimin Rotation Using a Correlation Matrix as the Input File

```
matrix data variables= Hugs Comps PerAd SocAc
 ProAd ComSt PhyHlp Encour Tutor
/file='f:\example1.txt'
/n=200
/contents=corr.

FACTOR
  /matrix=in(cor=*)
  /ANALYSIS Hugs Comps PerAd SocAc ProAd ComSt
 PhyHlp Encour Tutor
  /PRINT UNIVARIATE INITIAL EXTRACTION ROTATION
  /FORMAT SORT
  /PLOT EIGEN
  /CRITERIA FACTORS(2) ITERATE(25)
  /EXTRACTION ML
  /CRITERIA ITERATE(25) DELTA(0)
  /ROTATION OBLIMIN
  /METHOD=CORRELATION.
```

Note: The first five lines of code read in a text file containing the correlation matrix of measured variables. The remaining lines of code specify the factor analysis to be conducted on this correlation matrix.

statement, the first subcommand indicates that a matrix, rather than raw data, will be analyzed, and it indicates that this will be the correlation matrix read into SPSS in the prior commands. The next statement indicates the variables to be included in the analysis. The "print" subcommand indicates that univariate statistics (means, standard deviations, and sample size), initial communalities (squared multiple correlations—SMCs—for each item), pattern matrix and final communalities, and rotated and transformation matrices will be printed. The "sort" statement indicates that factor loadings will be sorted in descending order of magnitude. The "plot" statement indicates that eigenvalues will be plotted though, as we will see later, these eigenvalues are from the unreduced matrix. Thus, as noted earlier, a macro will have to be run in order to obtain the eigenvalues for the scree plot from the reduced (SMCs in the diagonal) matrix. On the first "criteria" line of the factor analysis code, two factors are specified as the number of factors to extract (this might often be adjusted across runs in order to compare models with different numbers of factors), and the maximum number of iterations is set for factor extraction. It should be noted that criteria commands in the SPSS factor analysis procedure always apply to the command line that follows the criteria statement. Next, the method of factor extraction is specified. In this case, it is ML. SPSS also supports other prevalent common factor model fitting procedures, such as IPA and NIPA, as well as the PCA model (for syntax to run IPA and NIPA factoring as well as PCA, see www.oup.com/us/exploratoryfactoranalysis). In the second criteria line, the maximum number of iterations for rotation is specified, and the delta value is specified as zero, thus indicating to SPSS that the direct quartimin rotation of the direct oblimin family should be conducted. In the next line, the type of rotation is specified as direct oblimin. SPSS also provides the promax (oblique) rotation and varimax (orthogonal) rotation as options. The "method" command indicates that a correlation matrix rather than a covariance matrix will be analyzed.

SAS. Table 5.9 presents the syntax used to conduct an ML EFA specifying two factors and using a Promax oblique rotation. Note that the first set of commands reads in the correlation matrix, specifying the variable names, location of the file, and nature of the entered data. This program reads in the correlation matrix

Table 5.9

SAS Syntax for a Maximum Likelihood EFA with Two Common Factors and a Promax Rotation Using a Correlation Matrix as the Input File

```
DATA d1 (TYPE=CORR);
   _TYPE_= 'CORR';
   INFILE 'f:\example1.txt' MISSOVER;
   INPUT Hugs Comps PerAd SocAc ProAd ComSt
 PhyHlp Encour Tutor;
RUN;

PROC FACTOR DATA=D1 NFACTORS=2 METHOD=ML
MAXITER=25 ROTATE=PROMAX SCREE RES;
RUN;
```

Note: The first 5 lines of code read in a text file containing the correlation matrix of measured variables. The remaining lines of code specify the factor analysis to be conducted on this correlation matrix.

directly. For syntax that runs on raw data to create the correlation matrix and complete the EFA, see the web site that accompanies the book (www.oup.com/us/exploratoryfactoranalysis). In SAS, the actual factor analysis is run using the FACTOR procedure. The PROC FACTOR statement includes options that specify the data set, the number of factors to retain, the method of factor extraction, maximum number of iterations, and the rotation method, and we also asked for a printout of the scree plot and the residual matrix (i.e., the matrix of differences between the correlations among measured variables predicted by the model and those actually observed in the sample). It should be noted that promax rotation (an oblique rotation) is specified in this syntax file because direct quartimin rotation is not available in SAS. SAS does provide options for conducting a Harris-Kaiser oblique rotation and varimax (orthogonal) rotation.

CEFA. Table 5.10 presents the script file used to conduct an ML EFA specifying two factors and using a direct quartimin oblique rotation (referred to as CF-Quartimax in CEFA). This script file can be created from drop-down menus or manually by simply typing the code into a text file that can be read by CEFA. The first two lines provide a title for the analysis and specify the input

Table 5.10
CEFA Script File for Conducting a Maximum Likelihood EFA with Two Common Factors and a Direct Quartimin (Labeled as CF-QUARTIMAX) Rotation

```
TITLE EFA Book Sample CEFA Analysis
DATAFILE   F:\EFA Book Sample CEFA Data File.inp
NUMFACT    2
DISCFUN    2
DISPMAT    1
MAXITER    50
DATADIST   1
ITERINFO   1
NUMDP      3
TYPROT     2
FNAME      CF-QUARTIMAX
ROWWT1     2
ROTEPS     0.000001
RNDSTARTS  0
USEORDMAT  0
NOMIN      0
SERMETH    1
TYPDER     4
STOP
```

data file (presented in table 5.11). The next few lines specify the factor analysis to be conducted. The third line (NUMFACT) indicates the number of factors that will be included in the model (two in the present case). The fourth line (DISCFUN) refers to the discrepancy function that will be used or, using more common EFA terminology, the model fitting or factor extraction procedure. Eight model fitting procedures are available in CEFA with a number corresponding to each. A value of 2 indicates that ML will be used. The DISPMAT (dispersion matrix) command indicates whether a correlation (value of 1) or covariance (value of 2) matrix will be analyzed. The MAXITER command indicates the maximum number of iterations for the model fitting procedure. The DATADIST command with a value of 1 indicates that the

Table 5.11
CEFA Data File for Maximum Likelihood EFA with Two Common Factors and a Direct Quartimin Rotation

```
200 9 (NObs, NVar)
1 (DataType)
0 (PRNGSeed)
1 (VarNames?)
hugs comp pers soc prof comm phlp encg tutr
0 (FacNames?)
0 (OrdMatFlag)
   1.000
   0.666 1.000
   0.150 0.247 1.000
   0.617 0.576 0.222 1.000
   0.541 0.510 0.081 0.409 1.000
   0.653 0.642 0.164 0.560 0.667 1.000
   0.473 0.425 0.091 0.338 0.734 0.596 1.000
   0.549 0.544 0.181 0.448 0.465 0.540 0.432 1.000
   0.566 0.488 0.120 0.349 0.754 0.672 0.718 0.412 1.000
```

data will be assumed to have a multivariate normal distribution. INTERINFO indicates whether information of each iteration should be saved (1=yes, 0=no). NUMDP indicates the number of decimal places that will be used in the output.

CEFA offers a wide range of different types of rotations. The next nine commands are used to specify different features of the rotation. TYPROT indicates whether only an orthogonal rotation will be used (1), only an oblique rotation will be used (2), or if a combination of two rotations will be used (3 = orthogonal-orthogonal, 4 = orthogonal-oblique). FNAME indicates the particular rotation that will be used in the first rotation. In the present example, a CF-QUARTIMAX has been specified, which corresponds to the direct quartimin rotation. ROWWT1 indicates if row standardization will be conducted on the first rotation (see the chapter 3 discussion of varimax rotation for a review of this issue). A value of 2 (Kaiser weights) is the most commonly used standardization, and it is specified in the present example.

FNAME2 and ROWWT2 specify the same properties for the second rotation if a second rotation is used. In the present example, because only a single rotation is being used, these commands can be omitted. The ROTEPS command indicates the rotation convergence criterion. The value in the current example is the default and is appropriate for most contexts. RNDSTARTS permits the researcher to investigate the possibility of local minima by examining the effects of different start values for the rotation. In the present example, this has been set to 0, which means no additional start values will be examined. CEFA is capable of conducting a target rotation in which the solution is rotated to some specified pattern of loadings (see chapter 3). The USEORDMAT indicates if a target pattern of loadings will be used (0 = no matrix provided, 1 = target matrix).

The remaining commands refer to advanced options in estimating model parameters and standard errors of these parameters. The present set of options specified in the example is the most commonly used. This set of options will be appropriate for the great majority of contexts in which researchers use EFA.

Table 5.11 presents the CEFA data file for our example. This file contains the actual data that CEFA will analyze and specifies some properties of the data. As with the script file, this file can be specified using the CEFA menus. The data to be analyzed can then be added to the end of the data file produced by CEFA. The first line in this file specifies the number of observations and the number of variables in the data. The second command line indicates if the data to be analyzed is in the form of a correlation or covariance matrix (1), raw data (2), a factor loading matrix (3), or raw data that will be used to compute polychoric correlations (4). The third command with a value of 0 indicates that, if random starting values are to be examined for a rotation, the system clock will be used to select the seed value to start this process. The fourth command indicates whether variable names will be provided (1) or no names will be provided (0). The fifth command indicates whether factor names will be provided (0 = no, 1 = yes). The sixth command indicates whether a target matrix will be provided (0 = no, 1 = yes). A value of 1 is selected only if the researcher is conducting a target rotation. The final set of lines provides the actual data that will be analyzed (in the present example, a correlation matrix).

Interpreting the EFA Results

The remaining sections of this chapter will focus on interpretation of results from a final EFA solution (once the number of factors and type of rotation are specified and the EFA is run). Our discussion will review each of the major types of information provided in an EFA solution. Because the available information from most EFA routines can be quite extensive (some of which goes beyond the scope this book), we will focus on only that information that will almost always be of central interest to researchers. Thus, we will discuss how to interpret communality estimates, factor loadings, interfactor correlations, and model fit. In the midst of these discussions, we will highlight how the information provided by EFA can be related to the two primary purposes for which it is used: construction of measurement instruments and identification of basic constructs. Within each section, the results from SPSS analyses will be used to illustrate our example. In almost all cases, presuming the same model fitting and rotation procedures have been selected, results in SAS or CEFA are identical or within rounding error of those based on SPSS (slight differences are not surprising given that different algorithms are used to implement iterative procedures). However, the different programs do sometimes use slightly different terminology to refer to particular aspects of the solution, and we will note such differences when they arise. Additionally, as already mentioned at several points in this chapter, some programs provide information not available in other programs. We will also note these differences during our presentation of the SPSS analyses. For full results from SAS and CEFA, readers can refer to www.oup.com/us/exploratoryfactoranalysis. The reader can also see output from a number of different types of models (including PCA and EFA models), fitting procedures, and rotations at www.oup.com/us/exploratoryfactoranalysis.

Communality Estimates. Communality estimates are very useful pieces information that researchers should always examine. Recall that each variable's communality is the variance in that measured variable explained by all the common factors in the model. The SPSS communalities are presented in table 5.12. The initial communalities in SPSS are the squared multiple correlations based on the other measured variables, and the final (extraction) communalities are the variance in items accounted for by the extracted factors. In our example, most of the items have moderate to high

Table 5.12

SPSS Communalities and Total Variance Explained for a Maximum Likelihood EFA with Two Common Factors and a Direct Quartimin Rotation

	Communalities	
	Initial	**Extraction**
Hugs	.612	.680
Comps	.567	.657
PerAd	.085	.075
SocAc	.472	.552
ProAd	.679	.765
ComSt	.653	.705
PhyHlp	.610	.695
Encour	.406	.435
Tutor	.679	.753

Extraction Method: Maximum Likelihood

Total Variance Explained

Factor	Initial Eigenvalues			Extraction Sums of Squared Loadings			Rotation Sums of Squared Loadings[a]
	Total	% of Variance	Cumulative %	Total	% of Variance	Cumulative %	Total
1	4.910	54.559	54.559	4.544	50.489	50.489	3.857
2	1.161	12.903	67.462	.773	8.586	59.075	3.600
3	.843	9.368	76.830				
4	.557	6.194	83.024				
5	.415	4.610	87.635				
6	.331	3.679	91.314				
7	.309	3.438	94.753				
8	.255	2.832	97.584				
9	.217	2.416	100.000				

Extraction Method: Maximum Likelihood.

[a] When factors are correlated, sums of squared load ngs cannot be added to obtain a total variance.

communalities, except for the Personal Advice item. Its low communality suggests that it is not loading on either of the extracted factors, and, thus,it is comparatively unrelated to the other items in the battery.

Such a finding might prompt a researcher to consider why the variable is not associated with the other variables in the battery. At a very general level, variables with low communalities could reflect measures with high levels of random error (perhaps the wording of the item is ambiguous). It could reflect that the variable of interest is not part of the same domain as other variables in the battery. For instance, given the nature of this variable in our example, one might speculate that even helpful people might be reticent to give Personal Advice, as such advice might not always be appreciated by the recipient. Instead, perhaps, this type of advice might be viewed as butting in or claiming some personal expertise that the respondent does not believe the advice giver merits. Finally, it could be that the personal advice item belongs to the domain of interest, but the sampling of the domain may not have been sufficiently broad to include other items that also tap into the same common factor as this item. That is, choosing whether or not to give personal advice might depend on different antecedents than choosing the other types of help addressed in this battery of items.

The same initial communalities as in SPSS appear in SAS under the label of Prior Communality Estimates: SMC and in CEFA as Noniterative Unique Variances, Communalities, and SMCs. In this part of the CEFA output, the SMCs are the initial communalities from SPSS. The final (extraction) communalities from SPSS appear in SAS as Communalities at each iteration (final communalities appearing at the last iteration) and in CEFA as MWL Unique Variances and Communalities.

The second panel in table 5.12 labeled Total Variance Explained presents the initial eigenvalues from the SPSS analysis. As was already noted, these eigenvalues are actually not the eigenvalues from the reduced matrix, but, instead, they are from the unreduced correlation matrix. Column five of this table presents the variance accounted for by the extracted factors prior to rotation. After rotation, the total amount of variance accounted for by the two factors will be the same, but this variance will be redistributed across the two factors. SAS reports the appropriate initial eigenvalues that correspond to the model fitting procedure being used and

Table 5.13
SPSS Unrotated Factor Loading Matrix and Goodness of Fit for a Maximum Likelihood EFA with Two Common Factors and a Direct Quartimin Rotation

Factor Matrix[a]

	Factor	
	1	**2**
ProAd	.834	−.261
ComSt	.834	.099
Tutor	.822	−.278
PhyHlp	.767	−.326
Hugs	.763	.314
Comps	.721	.370
Encour	.620	.226
SocAc	.607	.429
PerAd	.186	.200

Extraction Method: Maximum Likelihood.
[a] Two factors extracted. Four iterations required.

Goodness-of-fit Test

Chi-Square	**df**	**Sig.**
17.111	19	.582

CEFA does not report eigenvalues.

Factor Loadings. Table 5.13 presents the SPSS unrotated factor loading matrix and chi-square goodness-of-fit statistic. In SPSS, this is called the factor matrix. In SAS, the same matrix is called the factor-pattern matrix, and in CEFA, they are the MWL unrotated factor loadings. These loadings are usually not very useful in and of themselves, because they have not been rotated toward simple structure. The chi-square goodness-of-fit test is accurate for both the rotated and unrotated solutions, because rotation does not change the overall fit of the model to the data. As we have noted, the chi-square test is not in and of itself an especially useful test of model fit because it tests the extremely stringent hypothesis of perfect fit. However, in the present example, the test is not significant, thereby indicating that one cannot reject the null hypothesis that the model holds perfectly in the population. As noted earlier, the chi-square value can be used to derive descriptive fit indices that can be more useful such as RMSEA. SAS reports several indices in addition to the chi-square, but does not report RMSEA. CEFA reports several fit indices including RMSEA.

Table 5.14 presents the SPSS rotated factor loading matrix. Whenever an oblique rotation is used in SPSS, the rotated factor loading matrix is referred to as the pattern matrix. This is the matrix that should be the primary basis of interpretation. Its coefficients can be thought of as similar to standardized partial regression coefficients in which common factors are predicting measured variables (i.e., a factor loading is the standardized unit of increase in the measured variable for each standardized unit of increase in the common factor). Thus, these loadings represent impact of the factor on the item controlling for the impact of other factors in the model on the same item. It should be noted that, because factor loadings in an oblique rotation are similar to standardized partial regression coefficients, it is possible for these coefficients to be greater than 1.00 or less than −1.00, although they rarely go much beyond these values. Most factor analysis routines also report a second matrix of coefficients when an oblique rotation is used (in orthogonal rotations, only a single matrix of coefficients is reported). This second matrix is called the structure matrix in SPSS. It consists of the zero-order correlations between the factors and items (i.e., these associations do not control for impact of other factors). Researchers do not generally interpret this matrix. The factor correlation matrix, however, provides crucial information about the extent to which the factors are correlated with one another, and, thus, it is often of substantive interest to researchers.

In SAS, when oblique rotations are used, the rotated factor loadings (i.e., the standardized partial regression coefficients) are presented in the rotated factor-pattern matrix. The zero-order correlations between factors and measured variables appear in the factor structure matrix, and the interfactor correlations are simply labeled as such. In CEFA, the rotated factor loadings are in the CF-QUARTIMAX rotated factor matrix, and the interfactor correlations are labeled as factor correlations. CEFA does not report the structure coefficients. Both SAS and CEFA also report the standard errors for the factor loadings and interfactor correlations.

Table 5.14
SPSS Rotated Factor Loading Matrix, Structure Matrix, and Factor Correlation Matrix for a Maximum Likelihood EFA with Two Common Factors and a Direct Quartimin Rotation

Pattern Matrix[a]

	Factor	
	1	2
PhyHlp	.856	−.042
Tutor	.847	.036
ProAd	.839	.060
ComSt	.483	.467
SocAc	−.001	.744
Comps	.137	.726
Hugs	.221	.680
Encour	.209	.519
PerAd	−.067	.305

Extraction Method: Maximum Likelihood.
Rotation Method: Oblimin with Kaiser Normalization.
[a] Rotation converged in 9 iterations.

Structure Matrix

	Factor	
	1	2
ProAd	.873	.531
Tutor	.867	.511
PhyHlp	.833	.430
ComSt	.745	.738
Hugs	.603	.804
Comps	.544	.803
SocAc	.416	.743
Encour	.500	.637
PerAd	.104	.268

Extraction Method: Maximum Likelihood.
Rotation Method: Oblimin with Kaiser Normalization.

Factor Correlation Matrix

Factor	**1**	**2**
1	1.000	.561
2	.561	1.000

Extraction Method Maximum Likelihood.
Rotation Method: Oblimin with Kaiser Normalization.

As can be seen in table 5.14, the rotated factor loadings show a clean pattern, with hugs, compliments, social activities, and encouragement all loading 0.5 and above on the same factor (with loadings of 0.22 and below on the other factor). This factor could be interpreted as representing a willingness to provide psychological or social support. On the other hand, professional advice, physical help, and tutoring all load highly on the other factor, which might be interpreted as representing a preference to provide more material support. The remaining item, "communal studying," loads about equally (just below .5) on both factors. This makes sense given the nature of the factors. The communal nature of the experience may be related to other forms of psychological support (such as encouragement and social activities), but the studying itself may be related to other forms of material support (such as sharing professional advice or tutoring). Thus, the double-loading communal studying item is in no sense problematic. As noted in relation to the communalities, the personal advice item failed to load substantially on either factor. We noted several potential explanations for such a finding. Importantly, the occurrence of such items should always prompt researchers to carefully consider if its failure to load on any factors is indicative of poor item design, inadequate sampling of the domain of interest, or inappropriate inclusion of the variable within the battery.

What the researcher intends to accomplish with the research can help determine how the researcher uses the information from the rotated factor loadings. If the person was attempting to develop a scale to index each type of social support that people might offer, then she or he would probably opt to avoid use of the communal study item, because it loads equally on both factors. Thus, it is not a "pure" measure of either construct. Likewise, the personal advice item would not be included in either scale because it is a poor index of either construct.

However, if the researcher was more concerned with theory building than with developing a measure per se, then there is no necessary reason to shun the double-loading item. Indeed, its loading on both factors could be viewed as helping to clarify the nature of the factors at work. Any interpretation of the factors must not only account for the measured variables that load exclusively on one each factor. It must also make sense in terms of double-loading items. Hence, an interpretation that can account for single-loading and double-loading items would be more compelling than an interpretation that can only account for single-loading items. Also, if the researcher is looking for a setting in which a person can offer both types of support, the communal studying setting seems to be just such a setting. Thus, this item might alert the researcher to an especially interesting context in which to study the impact of each type of helping on student outcomes (such as relationships with fiends or family, academic outcomes, health, etc.).

Interfactor Correlations. One major benefit of using an oblique rotation is the information it provides about the possible correlations among factors. As noted in chapter 3, the oblique rotation does not require correlations among factors, but it does allow for correlations if those correlations would improve simple structure. In the current case, a substantial correlation between willingness to provide psychological and material support seems quite plausible. The correlation might suggest that some people like to help in both ways or that other common antecedents might motivate both types of support for others. SPSS does not report standard errors for factor correlations, but SAS and CEFA can report this information.

Summary. The two-factor solution appears to be quite interpretable. As intended in the simulation, the communalities are moderate to high for most of the measures. The low communality for one of the measures would alert the researcher to the divergence of that item from the other items intended to assess the constructs of interest. The factor loadings also make reasonable conceptual sense. The rotated factor loadings provide a clear picture of the relations among the items. The low-communality item fails to load on either rotated factor, and most of the other items load strongly on one factor and weakly or not at all on the other factor. One item has moderate loadings on both factors, but it is an item that conceptually relates to each of the factors, so it is easy to imagine that the item would load substantially on both. This double-loading item is not at all problematic for conceptualizing the factors and how the concepts relate to one another. Indeed, it can be helpful in that regard. However, if the intent of the research is to assess a set of items that could be used to create specific scales to assess each of the constructs, then the double-loading item might be dropped in favor of items that load clearly on one factor and little or not at all on the other. The interfactor correlation also suggests that the two types of social support under study are related to one another. This may raise interesting new research questions in terms of common antecedents that might motivate each type of helping.

Relating Factor Analysis Results to Scale Construction

To this point, we have treated EFA in a relatively generic manner, in which the analysis could be used for theory building, scale construction, or both. However, there are some specific senses in which the goal of scale construction, a common goal for EFA research, can be addressed by EFA results. We conclude our example by discussing two such senses: scale dimensionality and item selection.

Scale Dimensionality. Often, researchers construct a scale believing that it is unidimensional (i.e., that it assesses a single construct). However, the scale might, in actuality, be multidimensional. If researchers simply measure interitem consistency through a reliability measure, such as Cronbach's alpha, it is common for researchers to conclude that the scale is unidimensional if it has high reliability. Unfortunately, this is not always the case. It is possible for reliability to be relatively high even when scales are multidimensional in nature, particularly when factors are correlated (Cortina, 1993; John & Benet-Martínez, 2000). Also, sometimes researchers construct subscales thinking that they assess distinct constructs when, in fact, they do not. EFA can be used to detect such situations.

The question about the number of factors relates quite directly to the issue of scale dimensionality. If the items primarily tap into a single major factor, the EFA should reflect that fact. That is, the various number-of-factor procedures should suggest that a single-factor model is the most appropriate model for the data. In contrast, if the procedures suggest multiple factors, this indicates that the scale is likely to be multidimensional. Especially when each of the different procedures points toward a certain number of factors, this can provide a good basis to conceptualize the scale as reflecting a certain multidimensional structure. Also, the factor loadings help one to identify which items best capture each of the dimensions represented.

The interfactor correlations also provide one with information about the dimensionality and about the similarity or dissimilarity of the assessed constructs. When identified factors are highly correlated, one might also be more likely to think about whether the factors are related because they are influenced by a common higher-order factor. It is exactly in these situations involving hierarchical structure that different number-of-factor procedures might point to different numbers of factors. For example, a procedure like parallel analysis (that identifies factors larger than randomly occurring factors) might identify each of the lower-order factors, but a procedure like model fit (especially using a parsimony-influenced index like RMSEA) might be more likely to say that the higher-order factor is sufficient.

Item Selection. EFA also provides information about the extent to which each item is related to the underlying construct of interest. Items with low factor loadings are only weakly related to the construct of interest and thus are poor measures of that construct. Another valuable piece of information provided by factor analysis is that, when more than one construct is being assessed by a set of items (e.g., when a researcher is constructing two subscales), factor analysis provides the relations of items with all constructs. This will reveal items that might not be particularly "pure" measures of the underlying constructs (e.g., items that load substantially on more than one common factor, as was the case in our example for communal studying). Although such items might not be problematic in a theory-building context, it is generally best to avoid such measures when attempting to create separate indices of two related constructs.

Concluding Comments

In this chapter, we have reviewed several key aspects of implementing an EFA and interpreting its results. First, we discussed how three of the better performing number-of-factor procedures discussed in chapter 3 can be conducted with commonly used programs for conducting factor analysis (SPSS, SAS, CEFA). As we saw, none of the three programs was ideally suited to conduct all three procedures. However, with some supplementary programs and computations, it is possible to follow the recommendations provided in chapter 3. We then discussed the syntax used in each of the three focal programs to conduct an EFA and touched upon some of the procedural options available in each of the programs.

Finally, we examined the output provided by SPSS and discussed how it might be interpreted in the context of our example. At various points in our discussion of these results, we highlighted differences in the information provided by SPSS, SAS, and CEFA. Once again, none of the three programs was ideal in all respects regarding the information provided, but in each case, the program provided most of the crucial information, and additional information could be calculated or obtained through alternative means (such as macros or add-on programs).

References

Browne, M. W., & Cudeck, R. (1992). Alternative ways of assessing model fit. *Sociological Methods and Research, 21*, 230–258.

Cortina, J. M. (1993). What is coefficient alpha? An examination of theory and applications. *Journal of Applied Psychology, 78*, 98–104.

Fabrigar, L. R., Wegener, D. T., MacCallum, R. C., & Strahan, E. J. (1999). Evaluating the use of exploratory factor analysis in psychological research. *Psychological Methods, 4*, 272–299.

John, O. P., & Benet-Martínez, V. (2000). Measurement: Reliability, construct validation, and scale construction. In H. T. Reis & C. M. Judd (Eds.), *Handbook of research methods in social and personality psychology* (pp. 339–369). New York: Cambridge University Press.

Lorenzo-Seva, U., & Ferrando, P. J. (2006). FACTOR: A computer program to fit the exploratory factor analysis model. *Behavior Research Methods, 38*, 88–91.

MacCallum, R. C., Widaman, K. F., Preacher, K. J., & Hong, S. (2001). Sample size in factor analysis: The role of model error. *Multivariate Behavioral Research, 36*, 611–637.

MacCallum, R. C., Widaman, K. F., Zhang, S., & Hong, S. (1999). Sample size in factor analysis. *Psychological Methods, 4*, 84–99.

O'Connor, B. P. (2000). SPSS and SAS programs for determining the number of components using parallel analysis and Velicer's MAP test. *Behavior Research Methods, Instrumentation, and Computers, 32*, 396–402.

Steiger, J. H. & Lind, J. (1980, May). *Statistically based tests for the number of common factors.* Paper presented at the annual meeting of the Psychometric Society, Iowa City, IA.

Timmerman, M. E., & Lorenzo-Seva, U. (2011). Dimensionality assessment of ordered polytomous items with parallel analysis. *Psychological Methods, 16*, 209–220.

Wegener, D. T., & Fabrigar, L. R. (2004). Constructing and evaluating quantitative measures for social psychological research: Conceptual challenges and methodological solutions. In C. Sansone, C. C. Morf, & A. T. Panter (Eds.), *The SAGE handbook of methods in social psychology* (pp. 145–172). New York: Sage.

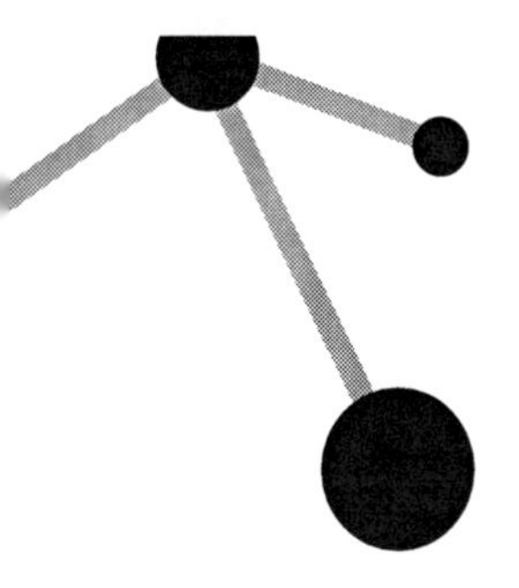

6

SUMMARY, CONCLUSIONS, AND RECOMMENDATIONS

THIS CHAPTER brings to a close our discussion of EFA. In this final chapter, we briefly review the key issues discussed in each of the preceding chapters and highlight some conclusions and recommendations that we think readily follow from our review. Obviously, one goal of this chapter is to provide readers with a summary of the key considerations that they should keep in mind whenever they are contemplating the use of EFA in their own research. However, we also see these conclusions and recommendations as important for anyone who is a consumer of research that makes use of EFA. Ultimately, a basic understanding of EFA is needed if a reader is to properly gauge the credibility of a given set of EFA findings.

Understanding the Common Factor Model

In chapter 1, we reviewed the basic premises of the common factor model, the general mathematical model on which EFA is based (Thurstone, 1935, 1947). As we discussed, this model is intended to explain the structure of correlations among a battery of measured variables. It postulates that a correlation exists between any two measured variables because the two variables are influenced by one or more of the same unobservable underlying constructs (i.e., common factors or latent variables). In addition, each measured variable is presumed to be influenced by a unique factor that represents both systematic influences and random effects that influence only that measured variable. We showed how this basic set of assumptions could be expressed both in mathematical form and in graphical form.

In discussing the common factor model, our hope has been to help researchers appreciate several aspects of this model. First, although the model can appear somewhat daunting when presented in its mathematical form, its underlying conceptual assumptions are comparatively simple and intuitive. Also, there are practical advantages to understanding the model. Although it is certainly possible in a mechanical sense to conduct an EFA without ever understanding the basic premises of the common factor model, many key decisions in implementing an EFA and interpreting its results are facilitated by understanding the model. Knowing when the model is appropriate or inappropriate to apply to a given data set, the conceptual distinctions between procedures such as EFA and PCA, and how to interpret the information provided by an EFA, such as factor loadings and communalities, are all aided by understanding the key premises of the common factor model.

Determining if Exploratory Factor Analysis is Appropriate

The central theme of chapter 2 was that prior to undertaking an EFA, there are key considerations that researchers need to take into account to determine if EFA is an appropriate analysis to conduct. First, researchers need to carefully consider the nature of the research questions they are examining. On the theoretical level, EFA is primarily used in the early phases of a research program when the goal is to identify the key constructs in the domain of interest. At the methodological level, EFA is generally used to aid in the development of measurement instruments by determining the dimensionality of a set of measured variables (e.g., a set of questions comprising a scale) and to determine the specific measured variables that best reflect the conceptual dimensions underlying the set of measured variables.

Although EFA can be used to address both theoretical (construct identification) and methodological (measurement construction) questions, these two uses of EFA are primarily a matter of emphasis rather than representing fundamentally different applications of the technique. Indeed, it is not uncommon for both goals to be addressed in the context of the same study. That is, a researcher may use EFA to reach conclusions about the number and nature of the key constructs in a domain of interest and then use the specific information for each measured variable provided by the analysis to select the subset of measures that best reflect each construct.

The second issue that researchers must consider is whether the data are suitable for EFA. This issue involves evaluating the soundness of the methods used to collect the data and the distributional properties of the measured variables to be analyzed. The impact of methodological choices on factor analysis is often underappreciated by researchers. Just as the results of an ANOVA must be evaluated in the context of the soundness of the experimental methods used to collect the data, so, too, are the results of EFA shaped by the soundness of the methods used to collect the data. An EFA is only as good as the data on which it is based. Thus, when designing studies that will use EFA or when evaluating results of EFAs conducted by others, it is important to attend to methodological features such how measured variables were selected (e.g., do the measures adequately represent the domain of interest?), how the measures were constructed (e.g., are the measures reliable and are assumptions of linear relations likely to be met?), and characteristics of the sample to which the measures are administered (e.g., is size of the sample adequate and is there sufficient variability in the sample on the constructs of interest?).

Third, assuming that factor analysis is appropriate, the researcher must then decide if an exploratory or confirmatory approach is most suitable to the research objectives. Many of the presumed differences between EFA and CFA are more a function of conventional practices than fundamental distinctions. Presumed advantages of CFA, such as the ability to compute model fit indices, to calculate standard errors for parameter estimates, and to conduct statistical tests of parameter estimates, are not actually advantages of CFA at all. The same practices can be applied to EFA when the EFA uses the same type of model fitting procedure as CFA. As noted in chapter 2 and later discussed in more detail in chapter 3, the most frequently used method of model fitting in CFA, maximum likelihood, can also be used in EFA. When it is used, information such as model fit, standard errors, and statistical tests can be computed for EFA (and some were originally developed for use in EFA rather than CFA).

Thus, the desire to formally evaluate model fit or conduct statistical tests of parameters should not serve as the basis to select a confirmatory approach over an exploratory approach. Instead, such a choice should be driven by the current state of theorizing and empirical evidence in the domain of interest. When the number and nature of common factors and their relations to measured variables are only poorly understood, EFA is generally the more sensible approach. When a single or only a few specific models can be plausibly advanced for a domain and/or there are very specific hypotheses to be tested regarding model parameters, CFA is appropriate.

Finally, assuming that an exploratory approach is adopted, the researcher must decide if an analysis based on the common factor model will be conducted or if an analysis based on the principal component model will be used. The decision about whether to use EFA or PCA has been among the most hotly debated issues in the factor analytic literature. Each approach has its advocates and its critics. For a variety of reasons outlined in chapter 2 (e.g., more realistic model assumptions, the specification of a falsifiable model, and the ability to evaluate model fit), our strong preference is for the common factor model used in EFA. At the very least, readers should recognize that at the conceptual level and the computational level, PCA is not the same as EFA. Principal components are mathematically different from common factors and thus, strictly speaking, PCA is not simply a type of factor analysis. Although some continue to discuss the relative merits of these two approaches, the mathematical distinctions between the two are not really debatable. On a practical level, there are instances in which EFA and PCA produce meaningfully different results, and EFA can provide information not available in PCA. Therefore, it is unwise for researchers to treat EFA and PCA as if they were interchangeable.

Decisions in Conducting Exploratory Factor Analysis

Assuming that a researcher has decided to conduct an EFA, that researcher must then consider a number of decisions in the implementation of the analysis. First, a model fitting (i.e., factor extraction) method must be selected. As indicated in chapter 3, our general preference is to use ML model fitting because of the additional information provided by this approach (e.g., model fit indices). That being said, there are contexts in which either IPA or NIPA factor analysis may be preferable, such as when assumptions of multivariate normality are severely violated or when there is reason to expect weak common factors that nonetheless may be of conceptual interest. Moreover, even when ML is used as the primary method of model fitting, it is a good idea to confirm that results also generally replicate using IPA or NIPA factor analysis. When results of the ML and principal axis procedures substantially diverge from one another, this occurrence can be indicative of problems with the model or the data (e.g., see Briggs & MacCallum, 2003; Widaman, 1993).

The second major challenge that a researcher must address in conducting EFA is to determine the appropriate number of common factors to specify. Perhaps no aspect of conducting an EFA is more challenging. Mistakes in determining the appropriate number of factors can lead to fundamental errors in conclusions. As discussed in chapter 3, some commonly used methods of determining the number of factors are highly problematic (e.g., the eigenvalues-greater-than-one rule) or are often misapplied (e.g., using eigenvalues from the unreduced correlation matrix for a scree test to determine the number of common factors). Other more defensible methods of determining the number of common factors are almost never used in practice, but they have considerable merit (e.g., parallel analysis, model fit). We recommend that researchers use three statistical procedures when determining the number of common factors: scree test of the eigenvalues from the reduced matrix, parallel analysis based on a common factor model approach (perhaps conducting both the procedure illustrated in chapter 5 that uses the reduced matrix with SMCs in the diagonal and the more recently developed procedure based on minimum rank factor analysis), and model fit using RMSEA. In addition to these procedures, researchers should also always carefully consider the conceptual plausibility/interpretability of the resulting model solution and, when possible, its stability across samples.

In discussing the number of factors issue in chapter 3, we attempted to stress several critical points. First, researchers need to keep in mind that this decision is as much a conceptual decision as it is a statistical decision. As such, it should not be based exclusively on mechanical rules. It requires informed judgment that balances statistical information with conceptual plausibility and utility. Hence, in some sense, for a given data set, there may sometimes be no unambiguous "true" number of common factors. Instead, there may be several defensible models for representing the data, and the choice among these models might well come down to the amount of error one is willing to accept in the model relative to the theoretical utility of the model (e.g., one model may account for the structure of correlations better than another model, but may be inferior to the other model with respect to its parsimony and/or interpretability).

A second point that we attempted to highlight is that not all of these factor number procedures are created equal. Some are problematic at the conceptual and/or practical level and are probably best avoided (e.g., the eigenvalues-greater-than-one rule). Third, even among the better-performing procedures for determining the number of factors, none of them is infallible. Some procedures certainly have a sensible underlying logic and function reasonably well in practice. However, none are perfect, and each approaches the number-of-factors question from a somewhat different perspective. In light of this fact, it is not surprising that these procedures sometimes suggest different conclusions. Thus, no single procedure trumps all others. Ultimately, the most defensible decisions are based on the full "configuration of results" rather than interpreting a single procedure in isolation. Thus, as a matter of course, we advocate examination of a number of the better-performing procedures, such as the scree test using eigenvalues from the reduced matrix, parallel analysis, and model fit, along with a careful consideration of the plausibility/interpretability of the solutions suggested by these procedures.

The final decision in implementing an EFA is to choose a factor-rotation procedure. In our discussion of rotation, we attempted to highlight several themes. First, we discussed the concept of simple structure and noted that it was frequently a sensible criterion to use in selecting a factor analytic solution. However, we noted that simple structure should not be expected in all contexts and, thus, researchers should not blindly adopt a simple structure rotation. Second, we noted that when simple structure is the goal of a factor analytic solution, the most fundamental choice researchers must make is between an orthogonal versus an oblique rotation. We believe that oblique rotations are generally more defensible than orthogonal rotations. Many of the presumed advantages of orthogonal rotation are based on misconceptions (e.g., that orthogonal rotations provide superior simple structure, that orthogonal rotations can make factors uncorrelated). In point of fact, successful oblique rotations are based on more realistic assumptions, generally provide solutions with comparable or superior simple structure, and provide more information than orthogonal rotations. For these reasons, we recommend that when simple structure is the goal of rotation, researchers should use one of the better-performing oblique rotations such as direct quartimin rotation (a member of the direct oblimin family of rotations).

Assumptions Underlying the Common Factor Model and Fitting Procedures

The primary focus of chapter 4 was to provide a more thorough discussion of assumptions underlying the common factor model and procedures for fitting the model. Chapter 4 addressed two key assumptions of the common factor model regarding the impact of common factors on measured variables. Specifically, the model assumes that common factors exert a causal influence on measured variables and that this influence is linear. Both assumptions are plausible in many contexts, but they should not be blindly accepted.

The issue of whether common factors cause measured variables requires a careful consideration of the nature of the measured variables being examined. We provided several examples in which the assumption of causal direction might not make sense. However, alternative conceptualizations, such as causal indicator models, also imply characteristics of the data and of the variables involved that can be questionable in many circumstances. With respect to assumptions of linearity, we noted that at least an interval level of measurement was a necessary but not sufficient requirement to meet this assumption. Thus, researchers should always carefully examine the nature of the response options or scoring system used for the measured variables to consider if an interval level or ratio scale of measurement is plausible. Beyond this basic precaution, evaluating assumptions of linearity are difficult, short of actually fitting nonlinear models and comparing their performance relative to the common factor model. However, when the linearity assumption is severely violated, it is worth noting that this problem can sometimes manifest itself in poor model fit or implausible parameter estimates. Thus, when such warning signs occur, nonlinear influences of factors might be one of several potential causes.

A final assumption made by some model fitting procedures (most notably ML) is that measured variables have a multivariate normal distribution. Fortunately, the ML-fitting procedure is relatively robust to violations of multivariate normality, but at severe levels distortions in results can occur. We noted that many traditional guidelines for evaluating distributional assumptions were problematic and suggested alternative guidelines as better standards against which to evaluate the severity of non-normality (i.e., an absolute value of skew of two or greater and an absolute value of kurtosis of seven or greater; West, Finch, & Curran, 1995). When data are severely non-normal, a variety of options exist for researchers, including data transformations, item parceling, and principal axis factor extraction.

Implementing and Interpreting EFA

The primary focus of chapter 5 was to provide readers with practical instructions about how to carry out an EFA consistent with the recommendations outlined in the prior chapters. Perhaps the one theme from this chapter that most merits commentary is the observation we made about the limitations of current factor analysis procedures in the major statistical packages (e.g., SPSS, SAS). These programs often fail to report valuable information (e.g., reporting of model fit indices is very limited for ML EFA procedures, and some programs do not report reduced matrix eigenvalues for common factor analyses). Also, many valuable factor analytic procedures are not offered (e.g., none of the programs conduct parallel analysis).

These limitations notwithstanding, we also attempted to highlight in chapter 5 that without too much additional effort on the part of researchers, it is possible to overcome many of these problems. SPSS and SAS programs have been written to perform some of the tasks not offered by the programs in their factor analytic options (e.g., O'Connor, 2000). Free programs exist that can further supplement the information provided by the major statistical programs (e.g., CEFA; Browne, Cudeck, Tateneni, & Mels, 2010; FACTOR; Lorenzo-Seva, & Ferrando, 2006). Thus, it is possible (and not too difficult) to conduct an EFA that is consistent with the recommendations advanced in this book.

Concluding Thoughts

More than 100 years after its initial development (Spearman, 1904), EFA continues to be an extremely valuable analytic method for researchers in a variety of substantive disciplines and a rich topic of inquiry for quantitative methodologists. Despite its longstanding centrality to and widespread use in many disciplines, perhaps few statistical methods have generated more controversy and have been more poorly understood. It is possible that one major reason for these difficulties is that few statistical procedures require more decisions and offer more procedural choices in implementing an analysis than EFA. On top of this, commonly available statistical packages have often made methodologically poor choices about the default options for EFA. Therefore, researchers lacking a firm conceptual and practical grounding in EFA can find the choices overwhelming and can unintentionally use poorly performing procedures. The goal of this book has been to provide a relatively complete but simple roadmap to help researchers navigate the challenges of conducting and interpreting EFA. Our belief is that when researchers have a basic understanding of the common factor model and the key procedural issues in its application, EFA can be an extremely useful tool.

Recommended Readings and Supplementary Programs

General References

Bollen, K. A. (1989). *Structural equations with latent variables.* New York: Wiley. (An excellent intermediate level overview of latent variable modeling. This book includes excellent discussions of the mathematical representation of the common factor model and various discussions of the conceptual issues related to the common factor model.)

Cudeck, R., & MacCallum, R. C. (Eds.). (2007). *Factor analysis at 100: Historical developments and future directions.* Mahwah, NJ: Erlbaum. (A recent survey of classic and contemporary topics in factor analysis by some of the leading contributors to the factor analysis literature.)

Gorsuch, R. L. (1983). *Factor analysis* (2nd ed.). Hillsdale, NJ: Erlbaum. (One of the most popular intermediate level texts of exploratory factor analysis. Some more important contemporary developments in EFA are obviously not covered, but the book is a good option for readers seeking a more extensive and mathematical coverage of factor analysis than the present book.)

Harman, H. H. (1976). *Modern factor analysis* (3rd ed.). Chicago; University of Chicago Press. (A classic text that is an excellent option for readers seeking a more advanced mathematical treatment of factor analysis.)

Tucker, L., & MacCallum, R. C. (1997). *Exploratory factor analysis.* Unpublished manuscript, University of North Carolina, Chapel Hill, NC. (An unpublished text on exploratory factor analysis that contains some excellent discussions of basic and advanced factor analysis topics. The first chapter is one of the best nonmathematical introductions to the common factor model that has been written. The book can be downloaded for free at: http://www.unc.edu/~rcm/book/factornew.htm)

Computer Software

Comprehensive Exploratory Factor Analysis (CEFA)

Browne, M. W., Cudeck, R., Tateneni, K., & Mels, G. (2010). CEFA: Comprehensive Exploratory Factor Analysis. Version 3.04 [Computer software and manual]. Retrieved from http://faculty.psy.ohio-state.edu/browne/ (This program provides researchers with additional information and a number of procedures not available in the standard factor analysis programs offered by SPSS and SAS. The program is available as a free download. A number of other useful programs, some of which are relevant to factor analysis, can also be found at the same site. Among these programs are FITMOD, which was discussed in chapter 5.)

Factor

Lorenzo-Seva, U., & Ferrando, P. J. (2006). FACTOR: A computer program to fit the exploratory factor analysis model. *Behavior Research Methods, 38,* 88–91. (This program is a general-purpose exploratory factor analysis program that includes a number of procedures not available in the standard factor analysis programs offered by SPSS and SAS. The program is available as a free download at http://psico.fcep.urv.es/utilitats/factor/index.html.)

SPSS and SAS Code for Number-of-Factors Procedures

O'Connor, B. P. (2000). SPSS and SAS programs for determining the number of components using parallel analysis and Velicer's MAP test. *Behavior Research*

Methods, Instrumentation, and Computers, 32, 396–402. (These SPSS and SAS programs allow for the performance of parallel analysis as well as the computation of eigenvalues from the reduced correlation matrix. The programs can be downloaded free of charge from: https://people.ok.ubc.ca/brioconn/nfactors/nfactors.html)

References

Briggs, N. E., & MacCallum, R. C. (2003). Recovery of weak common factors by maximum likelihood and ordinary least squares estimation. *Multivariate Behavioral Research, 38*, 25–56.

Browne, M. W., Cudeck, R., Tateneni, K., & Mels, G. (2010). CEFA: Comprehensive Exploratory Factor Analysis. Version 3.04 [Computer software and manual]. Retrieved from http://faculty.psy.ohio-state.edu/browne/

Lorenzo-Seva, U., & Ferrando, P. J. (2006). FACTOR: A computer program to fit the exploratory factor analysis model. *Behavior Research Methods, 38*, 88–91.

O'Connor, B. P. (2000). SPSS and SAS programs for determining the number of components using parallel analysis and Velicer's MAP test. *Behavior Research Methods, Instrumentation, and Computers, 32*, 396–402.

Spearman, C. (1904). General intelligence, objectively determined and measured. *American Journal of Psychology, 15*, 201–293.

Thurstone, L. L. (1935). *The vectors of mind.* Chicago: University of Chicago Press.

Thurstone, L. L. (1947). *Multiple factor analysis.* Chicago: University of Chicago Press.

West, S. G., Finch, J. F., & Curran, P. J. (1995). Structural equation models with nonnormal variables: Problems and remedies. In R. H. Hoyle (Ed.), *Structural equation modeling: Concepts, issues and applications* (pp. 56–75). Thousand Oaks, CA: Sage.

Widaman, K. F. (1993). Common factor analysis versus principal component analysis: Differential bias in representing model parameters? *Multivariate Behavioral Research, 28*, 263–311.

INDEX

CPSIA information can be obtained at www.ICGtesting.com
Printed in the USA
BVOW011223191012

303448BV00004B/9/P